Earthquake Interpretations

Earthquake Interpretations
A Manual For Reading Seismograms

Ruth B. Simon
Woodward-Clyde Consultants

William Kaufmann, Inc.
Los Altos, California

Dedicated to the Memory of
Don Tocher

ISBN 0-913232-81-5

10 9 8 7 6 5 4 3 2 1

Printed in the United States of America

Acknowledgments

The need for a book on the interpretation of seismograms exists even more now than in 1951 when Frank Neumann prepared the manual, *Principles Underlying the Interpretation of Seismograms*. The most recent revision (1966) of that publication, by the Government Printing Office, is still in use. This manual, and several other booklets, now out of print, by seismologists who attained eminence in the field, have contributed indirectly to this volume. I am especially indebted to Dr. Inge Lehmann of Denmark, and to the late Dr. Dean S. Carder for the example and inspiration they provided.

Professors John C. Hollister and Maurice W. Major gave my original publication, *Earthquake Interpretations*, the impetus needed for successful distribution by the Colorado School of Mines. The Publications Office of the Colorado School of Mines is gratefully acknowledged for permission to use previously published material.

Many colleagues at the U.S. Geological Survey contributed suggestions and records to add to this volume. Dr. Bruce A. Bolt of the University of California, Seismographic Station, Berkeley, provided records of the 1980 Livermore, California, earthquakes.

The Gray Milne Trust of the British Association for the Advancement of Science granted permission for reproduction of the Jeffreys-Bullen *Seismological Tables*. These are reproduced in their entirety in the Appendix.

My colleagues at Woodward-Clyde Consultants have helped this new volume to materialize. I especially appreciate the fine graphics assistance of David Hering. Critical review was provided by Dr. William U. Savage and technical editing was done by Janet L. Cluff.

Preface

"Listen to the earth, for it will teach you."

Seismology is still an infant of science, although earthquakes have no doubt occurred from near the beginning of the earth's existence. Despite its short life, seismology has made rapid advances. Increased knowledge of mathematics, physics, and geologic processes continue to advance earthquake studies. Much still remains to be done. The all-important question, *When will there be a devastating earthquake?* is unanswerable for the present. As earth science matures, and the study of seismology continues, the answers will come.

The purpose of this book is to help students interpret seismograms, knowing that every wiggle on the records can eventually be understood and explained. Several features of seismograms can be seen at a glance. The experienced observer knows immediately whether an earthquake is near or distant, from great depths within the earth or in the surficial crustal layers. Other features of a seismogram, perhaps not immediately apparent, teach the observer more about the earth by raising questions that can be answered only by more profound study and analysis of the seismogram. Computers now perform much of the detailed analysis necessary. Trained observers, however, are still needed to tell the computers what to do.

Earthquakes are like people in that no two are exactly the same. The complexity of the earth's structure causes the recording of a great variety of waves and wave types from shocks at different or even nearly the same locations. Therefore, a collection of seismograms to illustrate *typical* earthquakes is virtually impossible. The many factors influencing the appearance of a seismogram must be carefully noted and taken into account to achieve an accurate earthquake interpretation. The mechanism at the source of earth movement at the origin time of the earthquake, the path through and around the earth traveled by the seismic waves, the geologic structure along the path, and the geologic conditions beneath the recording station all affect the character of the seismogram in many presently accountable as well as unaccountable ways.

The instruments that record the earth's movement further complicate the picture by introducing factors, such as damping and filtering, that are not directly related to geologic structure or earth movements. The more experienced the reader is in interpreting from several sets of seismograms, the greater is the number of significant features that can be recognized on the records.

RUTH B. SIMON
Project Seismologist
Woodward-Clyde Consultants
Three Embarcadero Center
San Francisco, California 94111

Contents

Figures

Tables

Introduction

The following introductory material, reprinted by permission, and other selected instructional sections as noted, are modified from the *Manual of Seismological Observatory Practice* (Willmore, 1979). Using this material, which was prepared by a committee of the International Association of Seismology and Physics of the Earth's Interior, assures that this manual will be consistent with international practice in terms of methods of interpretation.

The Phenomenon of Earthquakes

All of us are aware of the human disasters that are sometimes caused by earthquakes. In the course of history, some of these have been of immense proportions, but the hazard is a continuous one in many parts of the world, and the few examples that follow will suffice to illustrate the magnitude of the problem:

1755	Lisbon, Portugal	60,000	dead
1783	Calabria, Italy	50,000	
1905	Kangra, India	20,000	
1908	Messina, Italy	83,000	
1915	Avezzano, Italy	30,000	
1920	Kansu, China	100,000	
1923	Kwanto, Japan	140,000	
1960	Agadir, Morocco	15,000	
1962	Qaevin, Iran	12,000	
1963	Skopje, Yugoslavia	1,500	
1968	Khorasan, Iran	13,000	
1976	Tangshan, China	500,000	
1980	El Asnam, Algeria	7,000	
1980	Southern Italy	3,000	

The rational response to any danger is to study the natural phenomenon that gives rise to it. In the case of earthquakes, the study has proceeded outwards from the centers of the most disastrous events. Thus, the earliest scientific studies were confined to the observation of gross (macroseismic) effects in the region surrounding the socalled *epicenter*, which was defined as the point on the surface of the ground at which the effects of the earthquake were most intense. At a later stage of development, it was found that earthquake waves could be detected instrumentally at great distances from the epicenter, and it was realized that the true source, or *focus*, was often at a very considerable depth within the earth.

The possibility of remote observation brought a new element of scientific detachment into the study of earthquake phenomena, by separating the contributing factors in the interaction of the natural event on a human environment. Sometimes, the instruments detect great earthquakes in sparsely inhabited parts of the earth, where little or no disturbance to human life is reported. At other times, great destruction and loss of life arise from the occurrence of a rather small earthquake in the immediate vicinity of a poorly constructed city. Nowadays, the instruments of the world network of seismograph stations have become so sensitive that the waves from minor earthquakes can be detected thousands of miles from the point of origin. The pattern of earthquake occurrence shows a great preponderance of energy release around the borders of the

Pacific Ocean (Figure A). This is seen to be markedly different from the pattern of earthquake effects on humanity, in which high population density combined with a high proportion of easily damaged buildings tends to concentrate the human toll in the Mediterranean basin, the Middle East and some parts of the Pacific coast.

The Objectives of Seismology

Recognizing the earthquake as a natural event that sometimes produces disastrous effects, the humanitarian may well place the alleviation of suffering at the head of his list of priorities. To achieve his main objective, he must combine an understanding of the natural phenomenon with the technological means of protecting people from harm. His operations must necessarily be influenced by local conditions, both in relation to the natural hazard and in respect of the resources that each community can deploy in its own defense.

One soon finds, however, that local problems cannot be considered in isolation. Earthquake hazards are related to soil conditions, geologic structure, and tectonic activity, which must be studied on a regional basis. A regional agency starts with a sphere of interest that may well extend across several national frontiers, but soon finds itself in need of data from other parts of the world. Eventually, one comes to realize that the structure is indivisible, and that organizations at all levels must cooperate within a worldwide framework. Logical exposition now requires the inversion of the human scale of priorities. The objectives and potentialities of the world network must be discussed first, so that the contribution that it can make to the solution of the regional problems can be understood. When the regional pattern has been revealed, we are in a position to consider the problems of the individual community.

In between the seismologist and the people who need protection from the effects of earthquakes, we find the structrual engineer and the earthquake engineer, who have the responsibility of ensuring that new structures have the proper degree of resistance to the earthquake hazards of their environment. Their interests unite with those of the seismologist in the estimation of the size and frequency of earthquakes in various parts of the world, and with architects, planners, and insurance companies in terms of social response to earthquake occurrences.

In view of the essential continuity of interest, it is a remarkable fact that interaction between the major groups remained at a comparatively low level until quite recently. Happily, the situation is improving, and numerous meetings are now taking place with the object of improving the understanding of the overall problem. The report of the Intergovernmental Conference on the Assessment and Mitigation of Earthquake Risk (UNESCO, 1976) makes an important contribution in this respect.

World Scientific Objectives

The locations of the Worldwide Standard Seismograph stations are mapped on Figure B. These stations are operated by the U.S. Geological Survey, which collects the data for the National Oceanic and Atmospheric Administration (NOAA), Environmental Data Service, at Boulder, Colorado, and for the International Seismological Centre (ISC), at Cambridge, England. Both agencies make available computerized lists of earthquakes. The NOAA records are also copied on 16 mm film. The ISC collects additional data from stations in Europe, Asia, the USSR, and neighboring countries; however, it does not itself operate any seismograph networks.

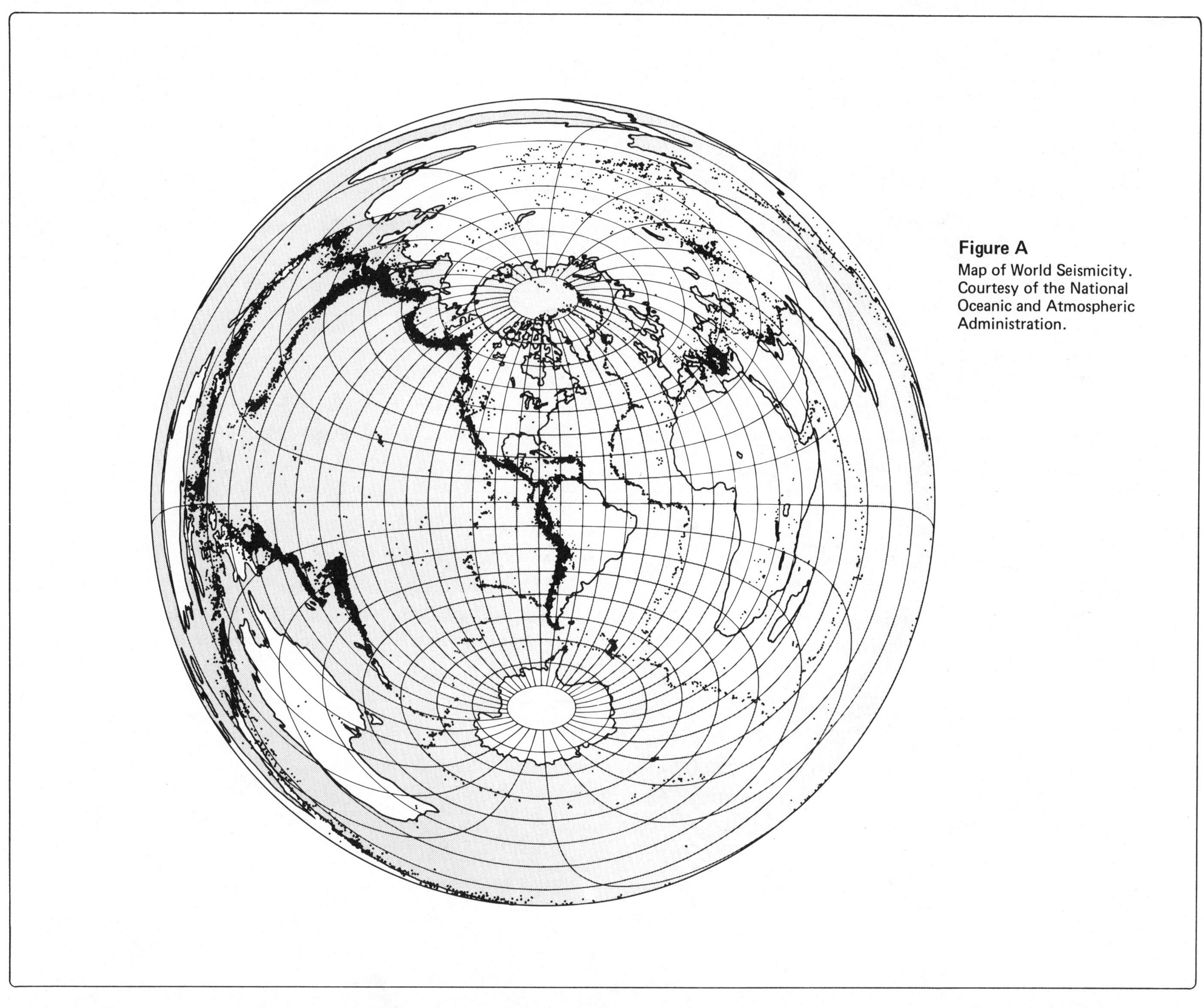

Figure A
Map of World Seismicity. Courtesy of the National Oceanic and Atmospheric Administration.

Map showing locations of the
Worldwide Standard Seismograph Stations

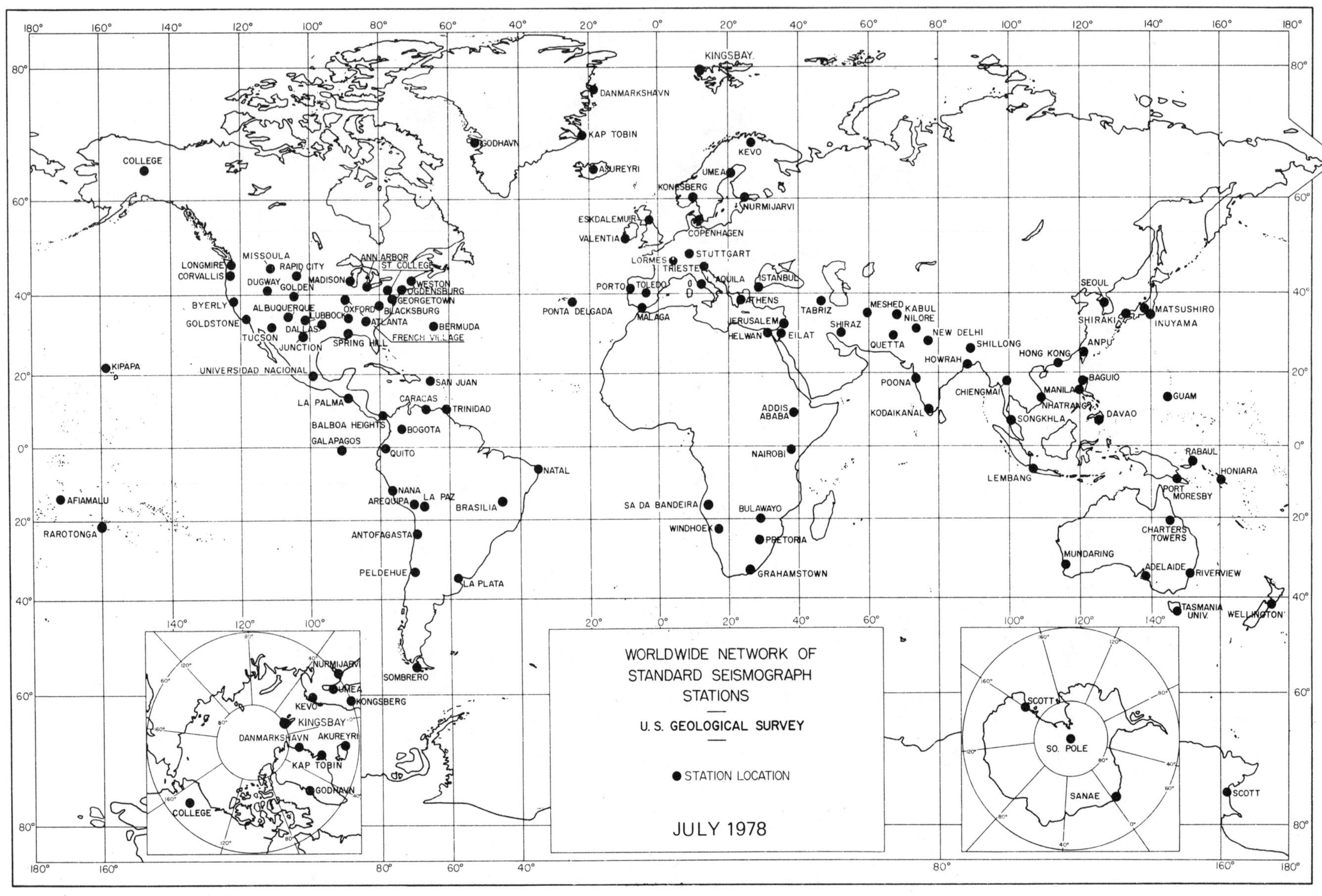

The permanent recording stations of the world network are collectively capable of responding to earthquake waves over a range of frequency extending from ten cycles per second down to those of the free oscillations of the whole earth, which have natural periods in excess of an hour. A few stations can detect the permanent distortion of the whole earth that results from the dislocation that occurs at the focus of a distant earthquake.

The waves that reach the station may have traveled through the interior of the earth, or may have been propagated as surface waves around its circumference, but in all cases, the information that can be derived from the record will be the time of occurrence, amplitude, and frequency of a visible disturbance. It is necessary to combine readings from a number of stations to determine the time, position, and magnitude of the source. When these parameters have been determined, the record of any one station can be interpreted in detail, and the information that it yields will relate to the passage of seismic energy along a variety of possible paths from the computed source to the known point of detection.

The instruments that are in active service have a wide range of sensitivity, and the environmental conditions range from those at the most isolated locations in the centers of continents, to those on oceanic islands and in the centers of cities. In consequence, the detection threshold varies over a range of more than 1000:1 between one station and another, and it may be more difficult to detect the passage of a seismic wave across a noisy site within 100 km of the source than it is to observe the same wave at a quiet station 10,000 km away. A few of the world's most sensitive stations are thereby producing a large proportion of all the available observations of small events, particularly those that occur outside the areas served by sensitive local networks. On the other hand, these high-sensitivity stations are easily deflected off scale by strong shocks, which are of particular importance for theoretical and practical investigation. It is therefore very desirable for such stations to include instruments of low and medium sensitivity in addition to their high-sensitivity equipment.

The main body of stations in the world network are producing readings of the waves from events that are substantially above the detection threshold of the most sensitive stations. The essential contribution of these stations is to provide simultaneous observations of medium-size events along numerous propagation paths through the earth. These interlocking observations provide a great deal of information about the mechanism of the event and about the internal structure of the earth. Uniform geographic coverage, wide range in frequency response, extended dynamic range, and the highest practicable degree of standardization between stations are the important requirements for this network. Many recommendations, made at international meetings, have been aimed at producing improvements in this respect.

It is particularly necessary to emphasize the contribution that aseismic countries can make, as a quiet, stable platform in a region of undisturbed geology is the chief requirement for long-range seismologic observations. The advantage of placing a seismographic station in a quiet position is comparable to that of siting an astronomical observatory far away from the lights of cities.

Short-range networks in seismic areas can also make a contribution to world seismology, by introducing additional precision into the timing and location of the events that occur within their radius of detection. The importance of this contribution is increasing as the world network becomes more sensitive, because events that a few

years ago were regarded as of purely local
significance are now being detected at great
distances.

We can therefore summarize the objectives of the
world network as the detection of seismic waves
and the location of their sources in space and
time. This basic information serves research in
the following ways:

- By combining the source data with detailed
  readings from selected stations, we investi-
  gate the propagation of waves through and
  over the earth, and thereby extend our knowledge of the structure of the earth along the
  propagation paths.

- The distribution of earthquake foci in space
  and time helps in understanding the origin of
  earthquakes, in delineating seismically
  active zones, and in studying geologic structure and dynamic processes in the crust and
  upper mantle of the earth.

Procedures For Beginners in Seismogram Interpretation

A seismogram is a recording of earth displacement as a function of advancing time. The recording may be at various magnifications and on various time scales, depending upon the speed of the motor that translates the recording medium. Magnification of actual ground motions by means of mechanical or optical amplification, with or without other electromagnetic or electronic amplification, is necessary for readable records.

Some seismograms, for example, those of the Worldwide Standard Seismograph Stations (WWSSS), are written on three-drum recorders that allow the three components of ground displacement to be recorded simultaneously. This is a great aid to the interpreter. It assures that the vertical, north-south, and east-west seismograms will record, all at the same time, not only earthquakes, but also calibration signals for amplification checks, hour and minute marks, and radio signals for time checks.

The purpose of initial seismogram interpretation is to determine the distance and direction of an earthquake epicenter from a recording station. The *epicenter* is a point on the surface of the earth directly above the *hypocenter*, which is the location of the focus of the earthquake. *Distance* (Δ) is defined as the arc of a great circle between the epicenter and the recording station, measured from an angle subtended at the center of the earth. It is expressed in degrees or in kilometers ($1° = 111.2$ km).

The various waves representing different wave types or wave groups of the same type arriving at a station over different paths are commonly called *phases*. The appearance of an earthquake record is very much influenced by the shape of the response curve of the seismograph. In fact, the seismograph works as a filter, amplifying only waves within a certain period range. The basic objective is to recognize the occurrence of an earthquake on the record, to describe the character of any given phase, and to determine the distance and time of origin correctly. There will be many confusing events, such as accidental disturbances on individual instruments, passing traffic or shocks in the building, or earthquakes too small to yield a reading (Willmore, 1979). The *time of origin* is the precise time the earthquake occurred. It is usually given in hours, minutes, seconds, and tenths of seconds in Greenwich Mean Time (GMT), now called Coordinated Universal Time (UTC). Because travel times of earthquake waves are computed from their origin times, timing must be accurate. Time signals used are emitted in the Coordinated Universal Time system and are broadcast in the United States over radio station WWV, Fort Collins, Colorado.

Two kinds of waves are recorded: *body waves* and *surface waves*.

<u>Body Waves</u> - Body waves are either (1) longitudinal, primary, "P" waves; or (2) transverse, secondary, "S" waves. P is the first wave seen on

a seismogram and represents a push-pull motion. P
waves are best recorded on the vertical components, although they may also be seen on horizontal component recordings. Of this type are also
pP, PP, PKP, PcP, and other multiples of P.

The pP is a wave reflected off the surface of the
earth and coming after P, indicating depth of
focus; the longer after P, the deeper the
hypocenter. The *depth* of an earthquake is the
distance in kilometers between the hypocenter
(focus) and the epicenter. PP indicates P waves
reflected once at the earth's surface. Phases
containing K have passed through the core of the
earth. Phases containing c have grazed the core
or are reflected from it. The *Manual of Seismological Observatory Practice* (Willmore, 1979)
summarizes the detailed nomenclature of seismic
phases for near earthquakes, distant shallow-focus
earthquakes, and deep-focus earthquakes. The
Manual also presents illustrations of how the
earthquake phases pass through the earth. (See
also *Seismological Tables* by Jeffreys and Bullen,
1967, in the Appendix to this volume.)

S is a later (secondary) arrival and represents a
transverse or shearing motion. S waves are best
recorded on horizontal components, although they
may often be seen as well on the vertical component recordings. Of this type are also sS, SS,
SKS, ScS, and other multiples of S.

Surface Waves - Surface waves are of two main
types: Rayleigh waves, LR, and Love waves, LQ.
Raleigh waves are longitudinal. Particle motion
is retrograde eliptical, traveling along the path
of propagation as:

LR usually can be seen on all three components.

Love waves are of the transverse or shear type.
Particles of these waves move with a shearing
motion along the path of propagation as:

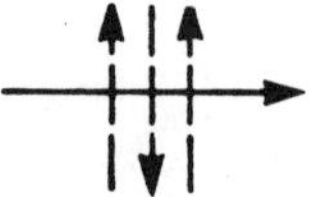

LQ can be seen only on the horizontal components.

Surface waves vibrate in many modes and have been
given different names by different investiga-
tors. Observed surface wave dispersion has been
summarized by Oliver (1962) and two sets of
dispersion curves relating phase and group velocities of Love and Raleigh waves as a function of
period have been reproduced in the *Manual of
Seismological Observatory Practice* (Willmore,
1979).

A first step in determining the distance of an
earthquake epicenter from a recording station is
to understand how to use a fundamental tool of
seismology, travel-time tables (Tables A and B)
and the curves constructed from these tables
(Figure C). The curves have been empirically
drawn from arrival times of seismic waves from
earthquakes recorded at stations at varying
distances from the epicenters. The Jeffreys-
Bullen Seismological Tables (Jeffreys and Bullen,
1967) have been used in this volume. They are
discussed in detail in the Appendix, where the
tables have been reproduced in their entirety for
the convenience of the reader. Other travel-time
tables have been proposed and are in use; however,
the Jeffreys-Bullen tables were the first and are
the most complete. They represent average,
worldwide data and give travel times and depth
corrections for all phases, whereas other tables

TABLE A

ARRIVAL TIME INTERVALS FOR NEAR
EARTHQUAKE PHASES
(Based on Jeffreys–Bullen Tables to 16°)

Δ km	Δ °	Time* Sn-Pn
8		1.0
9		1.1
10		1.2
11	0.1	1.3
12		1.4
13		1.5
14		1.7
15		1.8
16		1.9
17		2.0
18		2.1
19		2.2
20		2.4
21		2.5
22	0.2	2.6
23		2.7
24		2.8
25		3.0
26		3.1
27		3.2
28		3.3
29		3.4
30		3.5
31		3.7
32		3.8
33	0.3	3.9
34		4.0
35		4.1
36		4.3
37		4.4
38		4.5

Δ km	Δ °	Pn-O	Pg-Pn	Sn-Pn	Lg-Pg	Lg-Pn
39				4.6		
40				4.7		
41				4.8		
42				5.0		
43				5.1		
44	0.4			5.2		
45				5.4		
46				5.7		
47				5.9		
48				6.1		
49				6.4		
50				6.6		
51				6.8		
52				7.1		
53				7.3		
54				7.5		
55				7.8		
56	0.5			8.0		
	0.6	13.1		9.1		
	0.7	14.8		10.3		
	0.8	18.3	-0.5	11.4		
	0.9	19.7		12.5		
	1.0	21.1	-0.1	13.7		
	1.1	22.5	+0.4	14.7		
	1.2	23.9	0.8	15.9		
	1.3	25.4	1.2	16.9	18	19
	1.4	26.8	1.6	18.0	19	20
	1.5	28.2	2.0	19.2	20	22
	1.6	29.6	2.4	20.4	21	23
	1.7	31.0	2.7	21.5	22	25
	1.8	32.5	3.3	22.6	23	26

Δ °	Pn-O	Pg-Pn	Sn-Pn	Lg-Pg	Lg-Pn	Lg-O
1.9	34.0	3.9	23.8	24	28	
2.0	35.4	4.3	24.7	25	29	
2.1	36.8	4.6	25.9	26	31	
2.2	38.3	5.0	26.9	27	32	
2.3	39.8		28.1		34	1:14
2.4	41.2		29.2		35	1:16
2.5	42.6	6	30.4	31	37	1:20
2.6	44.0		31.5		39	1:23
2.7	45.5	7	32.6	34	41	1:26
2.8	46.9		33.7		42	1:29
2.9	48.3	8	34.8	36	44	1:32
3.0	49.7		35.9		46	1:36
3.1	51.1		36.9		48	1:39
3.2	52.5	9	38.1	41	50	1:42
3.3	53.9		39.2		51	1:45
3.4	55.4	10	40.3	42	52	1:47
3.5	56.8		41.5		55	1:52
3.6	58.2	11	42.6	46	57	1:55
3.7	59.7		43.7		59	1:59
3.8	1:01	12	44.8	48	1:00	2:01
3.9	1:02		45.9		1:02	2:04
4.0	1:04		47.1		1:04	2:08
4.1	1:05	13	48.1	53	1:06	2:11
4.2	1:07		49.3		1:08	2:15
4.3	1:08	14	50.3	55	1:09	2:17
4.4	1:10		51.4		1:11	2:21
4.5	1:11	15	52.8	58	1:13	2:24
4.6	1:12		53.8		1:15	2:27
4.7	1:14		54.7		1:17	2:31
4.8	1:15	16	55.9	1:02	1:18	2:33
4.9	1:17		57.1		1:20	2:37

Δ °	Pn-O	Pg-Pn	Sn-Pn	Lg-Pg	Lg-Pn	Lg-O
5.0	1:18		58.1		1:22	2:40
5.1	1:19	17	59.3	1:07	1:24	2:43
5.2	1:21		1:00		1:26	2:47
5.3	1:22	18	1:01	1:09	1:27	2:49
5.4	1:24		1:03		1:29	2:53
5.5	1:25	19	1:04	1:12	1:31	2:56
5.6	1:27		1:05		1:33	3:00
5.7	1:28	20	1:06	1:15	1:35	3:03
5.8	1:29		1:07		1:36	3:05
5.9	1:31		1:08		1:38	3:09
6.0	1:32	21	1:09	1:19	1:40	3:12
6.5	1:38	23		1:23	1:49	3:27
7.0	1:45	25		1:33	1:58	3:43
7.5	1:52	27		1:40	2:07	3:59
8.0	1:59	29		1:47	2:16	4:15
8.5	2:06				2:25	4:31
9.0	2:13				2:34	4:47
9.5	2:20				2:43	5:03
10.0	2:27	40		2:12	2:52	5:20
10.5	2:34	41		2:20	3:01	5:36
11.0	2:41	43		2:27	3:10	5:52
11.5	2:48	45		2:35	3:20	6:08
12.0	2:55	47		2:42	3:29	6:24
12.5	3:02	49		2:50	3:39	6:40
13.0	3:09	51		2:57	3:48	6:57
13.5	3:16	52		3:05	3:57	7:12
14.0	3:23	54		3:12	4:06	7:28
14.5	3:30	56		3:20	4:16	7:44
15.0	3:37	58		3:27	4:25	8:00
15.5	3:44	1:00		3:35	4:35	8:16
16.0	3:51	1:02		3:42	4:44	8:32

```
 *   Time is in seconds and tenths of seconds
 **  Time is in seconds and tenths of seconds, or minutes and seconds.
     Pg-Pn times are given in seconds for distances greater than 2.2°.
 Δ:  Distance from epicenter to station; 1° = 111.2 km
 Sn: Arrival time of S waves refracted below the Mohorovicic discontinuity
 Pn: Arrival time of P waves refracted below the Mohorovicic discontinuity
 O:  Origin time
 Pg: Arrival time of P wave in the upper crust
 Lg: Arrival time of crustal channel wave
```

TABLE B

ARRIVAL TIME INTERVALS FOR
DISTANT EARTHQUAKE PHASES
(Based on Jeffreys–Bullen Tables to 105°)

Δ°	P-O Min	P-O Sec	S-P Min	S-P Sec
0.0	0	(5.4)	0	3.8
0.5	0	10.5	0	7.6
1.0	0	17.7	0	13.1
1.5	0	24.8	0	18.7
2.0	0	32.0	0	24.1
2.5	0	39.1	0	29.7
3.0	0	46.3	0	35.2
3.5	0	53.4	0	39.9
4.0	1	00.5	0	46.4
4.5	1	07.6	0	51.9
5.0	1	14.7	0	57.4
5.5	1	21.7	1	03.0
6.0	1	28.7	1	08.5
6.5	1	35.8	1	13.9
7.0	1	42.8	1	19.3
7.5	1	49.8	1	24.8
8.0	1	56.7	1	30.3
8.5	2	03.7	1	35.8
9.0	2	10.6	1	41.3
9.5	2	17.5	1	46.8
10.0	2	24.4	1	52.2
11.0	2	38.1	2	02.9
12.0	2	51.6	2	13.7
13.0	3	05.0	2	24.4
14.0	3	18.1	2	35.2
15.0	3	31.2	2	45.7
16.0	3	44.1	2	56.1
17.0	3	56.7	3	06.6
18.0	4	09.2	3	16.8
19.0	4	21.5	3	27.0
20.0	4	32.5	3	38.1
21.0	4	42.9	3	47.1
22.0	4	52.9	3	55.8
23.0	5	02.8	4	04.0
24.0	5	12.5	4	11.8
25.0	5	22.2	4	19.1
26.0	5	31.6	4	26.2
27.0	5	40.8	4	33.4
28.0	5	49.9	4	40.4
29.0	5	58.8	4	47.5
30.0	6	07.7	4	54.5
31.0	6	16.6	5	01.4
32.0	6	25.4	5	08.2
33.0	6	34.1	5	15.1
34.0	6	42.7	5	22.0
35.0	6	51.3	5	28.9
36.0	6	59.8	5	35.8
37.0	7	08.2	5	42.8
38.0	7	16.6	5	49.6
39.0	7	24.9	5	56.5
40.0	7	33.2	6	03.2
41.0	7	41.5	6	09.8
42.0	7	49.7	6	16.4
43.0	7	57.9	6	22.8
44.0	8	06.0	6	29.3
45.0	8	14.0	6	35.7
46.0	8	22.0	6	42.0
47.0	8	29.8	6	48.4
48.0	8	37.7	6	54.7
49.0	8	45.4	7	01.0
50.0	8	53.1	7	07.2
51.0	9	00.7	7	13.4
52.0	9	08.2	7	19.7
53.0	9	15.7	7	25.8
54.0	9	23.1	7	32.0
55.0	9	30.4	7	38.1
56.0	9	37.6	7	44.3
57.0	9	44.8	7	50.3
58.0	9	51.8	7	56.4
59.0	9	58.8	8	02.5
60.0	10	05.7	8	08.5
61.0	10	12.5	8	14.5
62.0	10	19.2	8	20.5
63.0	10	25.9	8	26.3
64.0	10	32.4	8	32.3
65.0	10	38.9	8	38.1
66.0	10	45.3	8	43.9
67.0	10	51.6	8	49.7
68.0	10	57.9	8	55.4
69.0	11	04.1	9	01.1
70.0	11	10.2	9	06.8
71.0	11	16.3	9	12.3
72.0	11	22.2	9	18.0
73.0	11	28.2	9	23.4
74.0	11	34.0	9	28.8
75.0	11	39.8	9	34.2
76.0	11	45.5	9	39.5
77.0	11	51.2	9	44.7
78.0	11	56.7	9	50.0
79.0	12	02.2	9	55.1
80.0	12	07.6	10	00.2
81.0	12	12.9	10	05.3
82.0	12	18.1	10	10.4
83.0	12	23.2	10	15.4
84.0	12	28.3	10	20.3
85.0	12	33.3	10	25.2
86.0	12	38.2	10	30.0
87.0	12	43.1	10	34.7
88.0	12	47.9	10	39.3
89.0	12	52.7	10	43.8
90.0	12	57.4	10	48.3
91.0	13	02.1	10	52.5
92.0	13	06.7	10	56.7
93.0	13	11.3	11	00.9
94.0	13	15.8	11	05.0
95.0	13	20.4	11	08.9
96.0	13	24.9	11	12.9
97.0	13	29.5	11	16.8
98.0	13	34.0	11	20.7
99.0	13	38.5	11	24.6
100.0	13	43.1	11	28.4
101.0	13	47.6	11	32.3
102.0	13	52.1	11	36.1
103.0	13	56.5	11	40.1
104.0	14	00.9	11	44.0
105.0	14	05.3	11	47.9

Δ: Distance from epicenter to station
P: Arrival time of P
O: Origin time
S: Arrival time of S

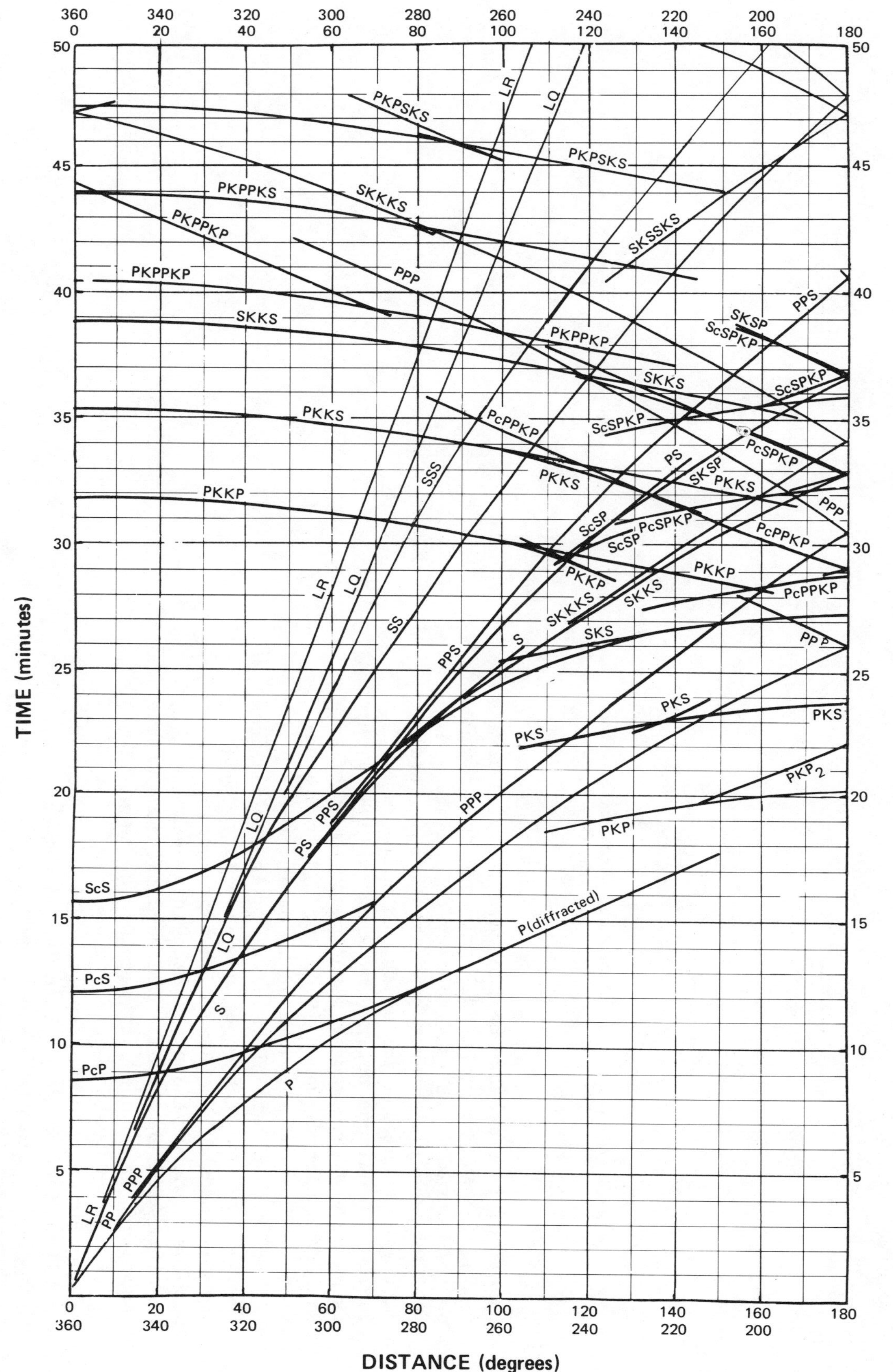

Figure C
Travel–time curves for earthquake phases.

are for P waves only, with slight variations from the Jeffreys-Bullen tables.

The *period* and *amplitude* of P waves are measured to compute earthquake magnitudes. The period is the first complete cycle of the wave. The amplitude is the height of the wave above its starting position.

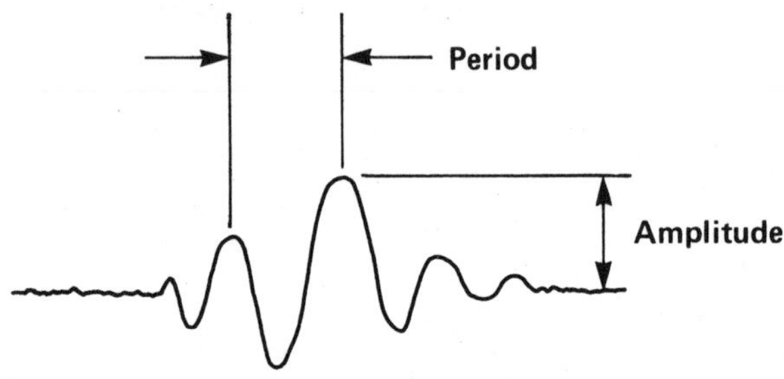

Magnitude is a measure of the relative size of an earthquake and is computed from measurements made on seismograms. It is frequency-dependent, and must be based on an average of stations. Three different magnitude scales are in common use. M_L is local magnitude defined by Richter (1935). M_L is usually determined from the measurement of maximum amplitude waves (P, S, or L_g) on a Wood-Anderson torsion seismometer recording for earthquakes at near distances. M_b is derived from the measurement of the period (1 to 10 seconds) and amplitude of the P wave on short-period records. These values are used in an empirical formula (Richter, 1958). The magnification of the instrument is a necessary part of the equation. M_s is a determination of magnitude made from the measurement of the amplitudes of 15 to 30 second period surface waves on long-period seismograms. These values are also used in an empirical formula (Richter, 1958). The magnification of the instrument is a necessary part of the equation for M_s also. Neither M_b nor M_s magnitudes are reliable if the earthquake focus is deeper than 33 km. The magnitude determination process is described in the *Manual of Seismological Observatory Practice* (Willmore, 1979).

Systematic Instructions for Interpretation

1. Equip yourself with a millimeter scale that has 1/2 mm gradations (1:100).

   Practice reading time on short-period records. Short-period seismograms of the WWSSS net record 15 minutes per line: 1 minute = 60 millimeters; 1 second = 1 millimeter.

   When you are familiar with this time scale, examine the long-period records. Long-period seismograms of the WWSSS net record 1 hour per line: 1 minute = 15 millimeters; 4 seconds = 1 millimeter.

   At many other stations, long-period records may be run on different scales such as 30 minutes per line. Then, 1 minute = 30 millimeters; 1 second = 2 millimeters.

2. Be sure of the time scale of the recording before beginning interpretation. Read accurately the hour, minute, second, and tenths of seconds. Check dates and times at the beginning and end of each record. Note clock corrections, if any. Stations other than WWSSS may have time marks that vary from WWV radio signals by a few seconds. This difference has to be added to or subtracted from each reading, depending upon whether the station clock is slow or fast.

3. Arrange the records so that you can see all the components at once: Z (vertical), N-S, E-W. Pick times of short-period body phases (P or PKP) on the short-period vertical seismogram. Find the same time on horizontal component recordings. Note, when possible, the directions of first motion on each, at the same time on each. The first motion, P, is either a compression (up) or a dilatation (down), and has a corresponding direction of motion either from or toward the source on horizontal records. Work on only one earthquake at a time.

4. Log phases giving the day, month, year, and e or i. An e indicates an emergent or gradual onset; an i indicates an impulsive phase, or sharp onset.

   After each e or i should be written the hour, minute, second, tenths of seconds, and direction of motion. The direction of motion is best obtained from impulsive arrivals when the precise instant of time is measurable on three components. It is better to omit the direction of motion than to log a questionable reading.

5. Measure the period, T, and the amplitude, A, of the P waves. The period, measured in seconds, is the first complete cycle of the wave. Peak-to-peak amplitude of this cycle is measured in millimeters. If a larger amplitude is seen in the next several cycles, the maximum amplitude should also be measured and logged with its time.

High frequency waves (T < 1 second) on short-
period records are usually from local sources.
These can be due to extraneous noise from
trucks, trains, sonic booms, or explosions.
High frequency waves with emergent onset are
usually not from an earthquake source.

6. After the earliest body phases have been
   picked, examine the long-period seismograms,
   using all three components at once. Often the
   P and PKP are seen here for larger earthquakes, although frequently PP is the first
   seen. Sometimes only S or SKS and later
   phases are seen.

   The most numerous seismic arrivals on long-
   period records are surface waves. These may
   be seen in earthquakes for which there is no P
   on short-period records.

   Read the arrival times of all surface waves at
   the beginning of the dispersed wave train.
   There are two types of dispersion. In normal
   dispersion, the train begins with longer
   period waves, with each successive cycle
   becoming shorter. Most of the records in this
   volume are of normally dispersed Love and
   Rayleigh waves. Inverse dispersion is the
   opposite; the train begins with shorter period
   waves, each cycle becoming longer. Both Love
   and Rayleigh waves can be inversely dispersed.

   ## Noise

   Wind or abrupt changes in barometric pressure
   or temperature are the most common causes of
   noise on long period seismograms. Noise can
   be distinguished from earthquake surface waves
   by the absence of dispersion. There is little
   or no change in period throughout the train.

Short-period noise on long-period records is
known as microseisms. These can have periods
from 2 to 20 seconds, but the most common
background noise has periods of 4 to 8
seconds. Microseisms often appear as storms
that have the same period of high amplitude
waves lasting many hours, or days, and are
seen on both short and long-period seismograms. The cause of this continuous activity
of the earth is related to ocean waves
impinging on continental coasts.

7. Direction of motion can be obtained from
   surface waves. The particle motion of
   Rayleigh waves is retrograde elliptical along
   the path of propagation; therefore, the
   particle motion at the time of arrival of the
   first Rayleigh wave peak on the long-period
   vertical seismogram coming from the source is
   up (↑) at a selected time. This same time on
   each of the horizontal component recordings
   will tell from which directions the waves are
   coming. On the N-S seismogram, if N is up
   (↑), and the next peak after the time of the
   vertical peak is also up, the particles are
   moving toward the south, or from the north
   (diagram a). Conversely, if the next observation after the vertical peak time is a
   trough, the motion is toward the north, or
   from the south (diagram b). The same process
   is followed with the vertical and the E-W
   components. Superimposing the records on a
   light table is helpful.

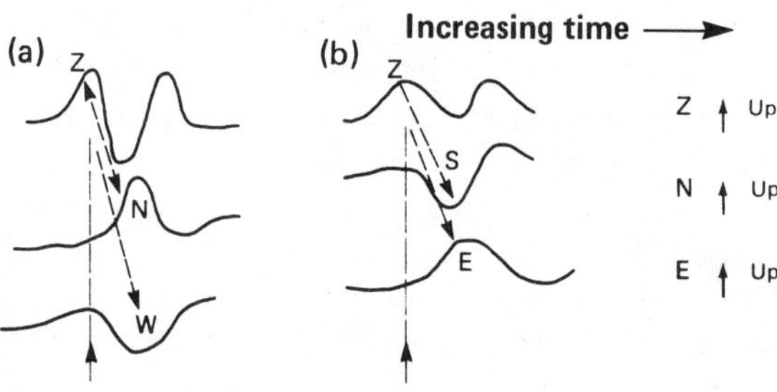

Comparison of amplitudes of Rayleigh waves on
the two horizontal components may show one
higher than the other. If higher on the N-S
component, the earthquake may be more north or
south. If higher on the E-W, the earthquake
may be more east or west. There is a
difference in locations at NW, NNW, or WNW
azimuths. Some observers make plots of
amplitudes on each component to measure an
angle of azimuth. However, this assumes that
the components are exactly matched in magnifi-
cation--not always a safe assumption.

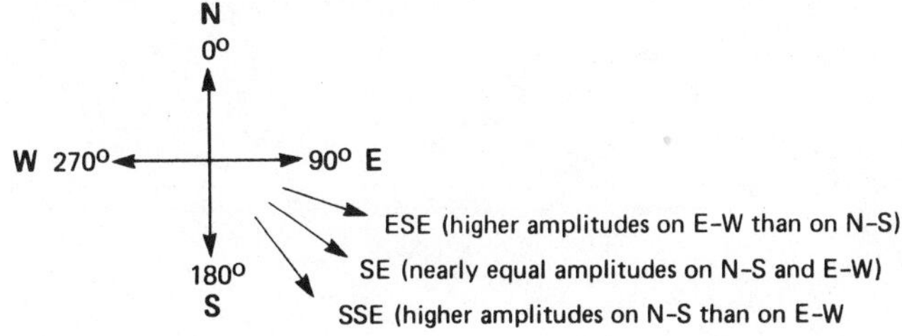

In some earthquakes where Love waves are
prominent, it is possible to check the azimuth
by superimposing one horizontal record on the
other. Care must be taken here because of
instrumental phase shifts and changing periods
observed. It is difficult to sort out Love
and Rayleigh waves later in the surface wave
train. The following azimuths may be seen
when the direction on each component is the
same as in the illustrations: Z↑, N↑, E↑.

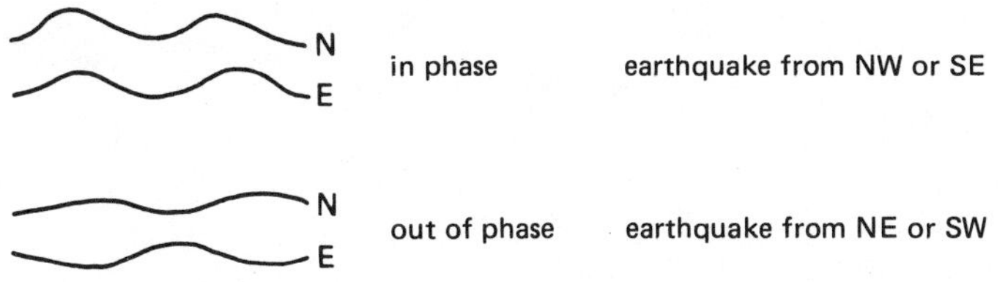

When possible, check azimuth from both Love
and Rayleigh waves. Where the Rayleigh waves
are of comparable amplitudes on all three
components, if the Love wave appears large on
the E-W component, the earthquake is much more
from the north or south.

Compare the direction of motion obtained from
P and LR. If there is disagreement, there may
be two earthquakes or a mistake in reading the
azimuth of P. Surface wave directions are
more dependable because the time span of
observation is longer. Azimuths obtained from
peaks of surface wave trains can be read for
one or more hours, but routinely should be
determined in the first 2 to 5 minutes of the
train.

8. When the listing of body phases and surface
    waves is complete, (designated only e, i, LQ,
    and LR) for one earthquake, it may then be
    convenient to prepare a paper tape. For this
    purpose, adding-machine tape is useful;
    otherwise, any strip 12 inches or longer will
    do. Mark off 5-minute intervals from the
    travel-time curve (Figure C). Insert at the
    appropriate times the arrivals you have
    listed. Then slide the tape across the
    travel-time curve matching your first pick
    with P or PKP, fitting S or SKS, LQ, and LR to
    find an approximate distance ($\Delta°$). Once you
    have ascertained which picks are P and S by
    this method, subtract P time from S time and
    consult the table of S-P (Table A or B) for a
    more accurate distance determination.

For deep earthquakes, phase arrival times will
be earlier than those shown on the surface
focus travel-time curve (Figure C). This
curve is usable for earthquakes as deep as
150 km. Deeper shocks require the use of the

Jeffreys-Bullen *Seismological Tables* in the
Appendix. Depth corrections are subtracted
from surface focus arrival times. The clue to
a shock occurring at great depth is the
absence of surface waves or extremely high
amplitudes of body waves relative to the
amplitudes of surface waves.

9. With the preliminary distance and azimuth, the
    location of an earthquake can now be approxi-
    mated on a globe. Most convenient is a globe
    that can be freely rotated in a stand
    calibrated in degrees (or a paper tape can be
    calibrated for the scale of any globe).

    By aligning the station at the correct azimuth
    with the presumed epicenter along the great
    circle, the location of the earthquake can be
    obtained with fair accuracy.

10. The *Manual of Seismological Observatory
    Practice* (Willmore, 1979) can be considered a
    companion volume to the record collection and
    brief notes assembled here. This *Manual*
    concisely presents more complete and technical
    seismological information that will be of
    great value to the serious student. Direc-
    tions for computing magnitudes are found in
    this *Manual*, as well as information on how to
    transmit your readings and other information
    to the appropriate agency.

The seismograms in this collection are arranged in order of increasing distance of the recording station from the earthquake epicenter, from 18 kilometers to 700° or 77,840 kilometers.

The components are placed in the same order in each illustration:

- Vertical (Z↑), topmost; motion up on the record indicates ground motion up.

- North-south (N↑), in the center; motion up on the record indicates ground motion north.

- East-west (E↑), at the bottom; motion up on the record indicates ground motion east.

The majority of the records are from two stations (Figure D): Cecil Green Geophysical Observatory of Colorado School of Mines, Bergen Park, Colorado (GOL) of the Worldwide Standard Seismograph Stations (WWSSS) network, and Lamont-Doherty Geological Observatory, Palisades, New York (PAL). Also included are recordings from Wood-Anderson torsion (horizontal) seismometers; and high-gain, high-frequency, closely spaced network recordings written on helicorders, develocorders, or magnetic tape (analog playback reproduced) with varying notations of time and consisting of vertical components only. Each trace has the component and time appropriately marked.

The seismograms are from two standard types of instrument systems. The short-period instruments (Benioff) have seismometer periods of 1.0 second and galvanometer periods of 0.75 second. These operate at a gain of 200,000 October through April, and 400,000 May through September for GOL records. The short-period records from all other stations are one Hertz with different electronic filters. Wood-Anderson torsion seismometers are 0.8 Hertz and have a standard magnification of 2800. The long-period instruments (Sprengnether, Press-Ewing, and Columbia) have seismometer periods of 30 seconds and galvanometer periods of 100 seconds, and operate at a gain of 1500 for both PAL and GOL records. Two records are included from Bergen Park strainmeters, which record the longest-period earth waves so far observed, with periods ranging from 30 seconds to one hour, as well as superimposed earth-tides with periods of 12 hours.

Comments about each earthquake in this collection are brief to encourage students to note for themselves the similarities, differences, and unusual features of each set of recordings. The phases are identified only when computed arrival times agree within one or two seconds of the observed arrival times. The Jeffreys-Bullen travel time tables have been used throughout. Earthquakes are included from similar distances, but different azimuths, or from similar distances, but different depths of focus, for comparison.

Figure D

Map showing locations of stations from
which the earthquake records were obtained

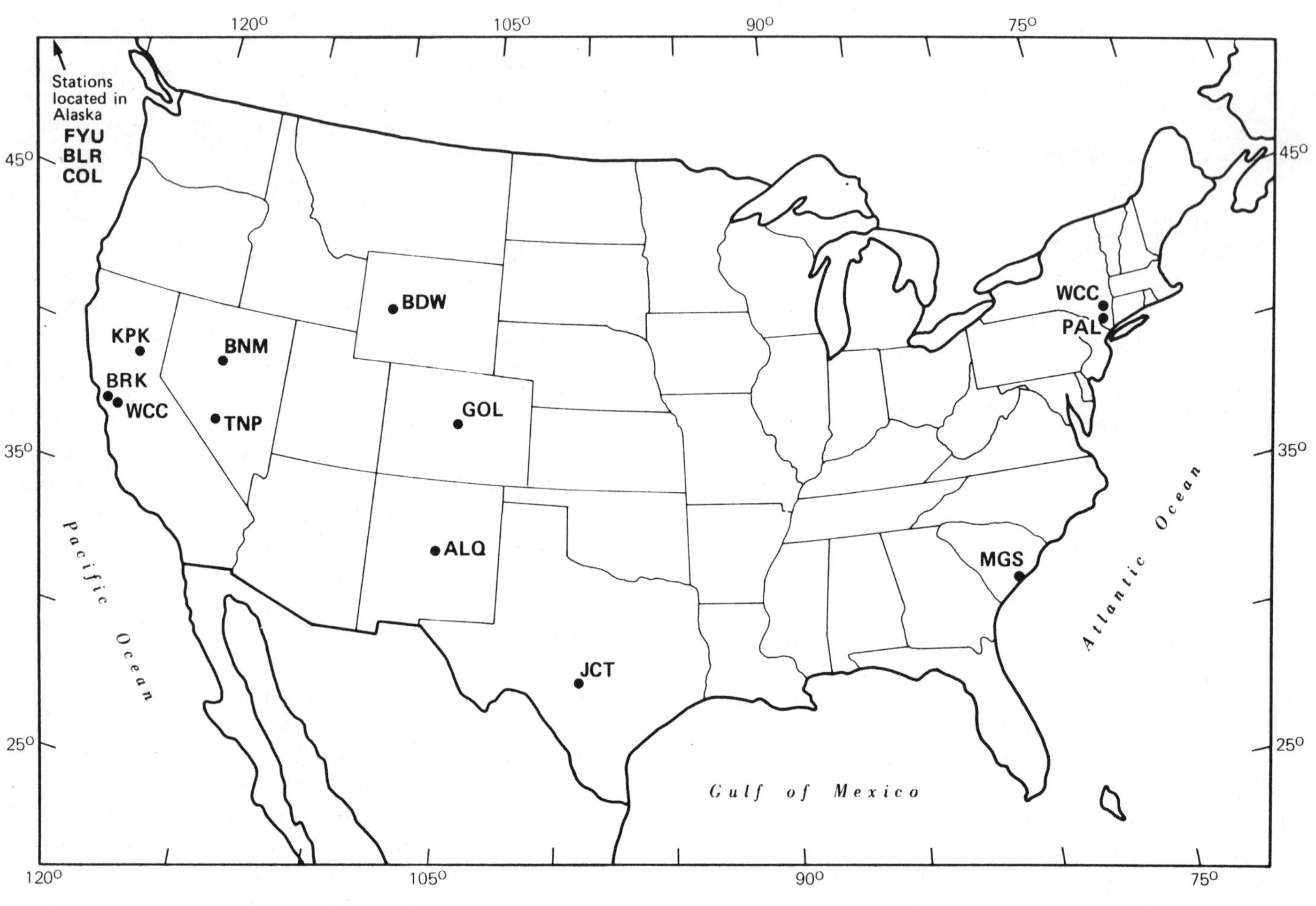

References

Short-period records are full size, one minute equals 60 mm, except for two (Figures 23 and 25) that are reduced 50 percent, one minute equals 30 mm. Long-period records are either full size; one minute equals 15 mm, or reduced 40 percent: one minute equals 6 mm.

Epicentral locations, origin times, focal depths, and magnitude determinations have been provided by the U.S. Geological Survey, Office of Earthquake Studies, Seismological Summaries, unless otherwise indicated. The epicentral distances have been computed from each station used.

Principles of operation of standard seismograph systems, as well as detailed descriptions of seismic wave types and their notation can be found in the *Manual of Seismological Observatory Practice* by Willmore (1979) and in other sources listed in the Selected Bibliography.

Jeffreys, H., and Bullen, K. E., 1967, *Seismological Tables*, London, British Association for the Advancement of Science, Gray Milne Trust, 50 p.

Oliver, J., 1962, A summary of observed seismic surface wave dispersion, *Bulletin of the Seismological Society of America*, v. 52, no. 1, p. 81-86.

Richter, C. F., 1935, An instrumental earthquake magnitude scale, *Bulletin of the Seismological Society of America*, v. 25, no. 1, p. 1-32.

Richter, C. F., 1958, *Elementary Seismology*, San Francisco, W. H. Freeman, 768 p.

UNESCO, 1976, *Intergovernmental Conference on the Assessment and Mitigation of Earthquake Risk*, Paris, UNESCO, 50 p.

Willmore, P. L., ed., 1979, *Manual of Seismological Observatory Practice*, World Data Center A for Solid-Earth Geophysics, Report SE-20, Boulder, National Oceanic and Atmospheric Administration, Environmental Data and Information Service.

Figure	Distance	Station	Type of Instrument	Location of Earthquake	Page
1A, 1B	18 km to 40 km	WCC	Short period	Oakland, New Jersey	23
2A	25 km	GOL	Short period	Loveland Pass, Colorado	26
2B	28 km	GOL	Short period	Cabin Creek, Colorado	
3A	33 km	BRK	Strong-motion torsion	Livermore, California	28
3B	47 km	BRK	Strong-motion torsion	Livermore, California	
4	40 km	GOL	Short period	Derby (Denver), Colorado	30
5	75 km	GOL	Short period	Castle Rock, Colorado	32
6A, 6B	84 km to 190 km	WCC	Short period	Mammoth Lakes, California	35
7	140 km	GOL	Short period	Aspen, Colorado	38
8	200 km	GOL	Short period	Somerset, Colorado	40
9	220 km	COL	Short period	Alaska	42
10	285 km	GOL	Short period	Silverton, Colorado	44
11	320 km	JCT	Short period	West Texas	46
12A	330 km	GOL	Short period	Dulce, New Mexico	48
12B	330 km	GOL	Long period	Gasbuggy Underground Nuclear Test, New Mexico	
13A	366 km	TNP	Short period	Central California	50
13B	400 km	KPK	Short period	Central California	
13C	610 km	BMN	Short period	Central California	
14	455 km	JCT	Short period	New Mexico	52
15	456 km	GOL	Short period	Monument Butte, Wyoming	54
16	5.3° (600 km)	GOL	Long period	Richfield, Utah	56
17	5.9°	GOL	Short period	Eastern Idaho	58
18	6°	GOL	Short period	Winner, South Dakota	60
19	8°	MGS	Short period	North Atlantic Ocean	62
20	9°	GOL	Short period	Yucca Flats, Nevada	64
21A, 21B	9°	GOL	Long period	Yucca Flats, Nevada	66
22	11°	GOL	Long period and Wood-Anderson strong-motion	California-Mexico Border (Imperial Valley)	68
23	13°	GOL	Short period	Gulf of California	70
24	13°	GOL	Long period	Gulf of California	72
25	16°	GOL	Short period	Offshore Northern California	74

26	16°	GOL	Long period	Offshore Northern California	76
27A	19°	GOL	Long period	Revilla Gigedo Islands	78
27B	23°	GOL	Long period	Vera Cruz, Mexico	
28A	29.4°	PAL	Long period	North Atlantic Ocean	80
28B	30.2°	PAL	Long period	El Salvador	
29A	30°	PAL	Long period	Offshore Nicaragua	82
29B	35.5°	PAL	Long period	Offshore Central America	
30A	34°	GOL	Long period	Gulf of Alaska	84
30B	34°	GOL	Short period	Gulf of Alaska	
31A	39.2°	PAL	Long period	Offshore Northern California	86
31B	44.5°	PAL	Long period	Southeastern Alaska	
32	43°	GOL	Long period	Northern Colombia	88
33	51.3°	PAL	Long period	Kodiak Island, Alaska	90
34	54°	GOL	Long period	North of Svalbard	92
35A	59.8°	PAL	Long period	Northern Chile	94
35B	70.8°	PAL	Long period	Greece	
36A	78°	GOL	Short period	Argentina	96
36B	78°	GOL	Long period	Argentina	
37	81°	GOL	Long period	Honshu, Japan	98
38	85.5°	GOL	Long period	Albania	100
39	80° to 89°	KPK,TNP BMN,ALQ BDW	Short period	Tonga Islands	102
40A	91.5°	BLR	Short period	Offshore Peru	104
40B	93°	FYU	Short period	Offshore Peru	
41	92.2°	GOL	Long period	South of Fiji Islands	106
42A	99.5°	GOL	Long period	Solomon Islands	108
42B	100°	GOL	E-W Strainmeter	New Hebrides Islands	
43	108°	GOL	Long period	East New Guinea Region	110
44	122°	GOL	Long period	Bombay, India	112
45	135°	GOL	Short period	Java	114
46	139.6°	PAL	Long period	Banda Sea	116
47	700°	GOL	N-S Strainmeter	Southern Alaska	118

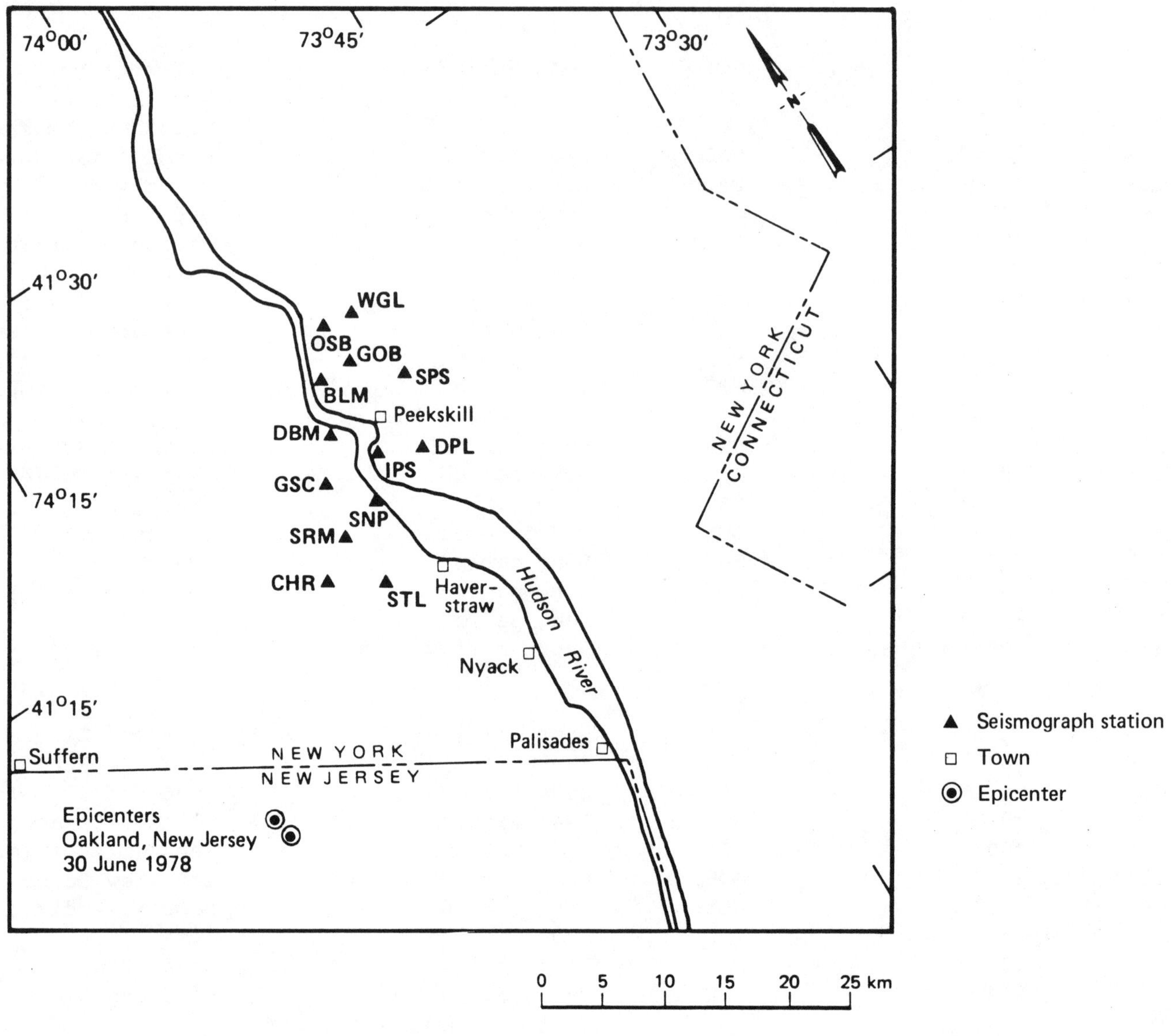

Map showing locations of stations in WCC array. The stations were operated by Woodward-Clyde Consultants, Clifton, New Jersey, for Consolidated Edison Company of New York.

Figure 1

	A	B
DISTANCE:	18 km to 40 km	18 to 40 km
RECORDING STATION:	WCC array; see listing in Comments	WCC array; see listing in Comments
INSTRUMENTS:	SP (2 Hz) Z, playback of magnetic tape recording	SP (2 Hz) Z, playback of magnetic tape recording
LOCATION:	New Jersey (Oakland vicinity)	New Jersey (Oakland vicinity)
COORDINATES:	41.08 N, 74.20 W	41.08 N, 74.20 W
DATE:	30 June 1978	30 June 1978
TIME:	20 13 43.5 UTC	22 39 49.7 UTC
DEPTH:	5 km	5 km
MAGNITUDE:	2.9 M_{bLg} (PAL)	2.2 M_{bLg} (PAL)

COMMENTS:

A:

Station (see map)	Distance (km)	Station (see map)	Distance (km)
CHR	18.4	DBM	30.0
STL	20.4	DPL	30.9
SRM	22.4	BLM	34.0
SNP	26.0	SPS	35.6
GSC	26.0	GOB	35.8
IPS	29.6	OSB	38.3
		WGL	39.5

B:

Station (see map)	Distance (km)	Station (see map)	Distance (km)
CHR	19.4	DBM	31.0
STL	21.1	DPL	31.5
SRM	23.3	BLM	35.1
SNP	26.8	SPS	--
GSC	27.0	GOB	36.8
IPS	30.0	OSB	39.4
		WGL	40.5

This suite of playback recordings was telemetered from the Indian Point, New York network operated by Woodward-Clyde Consultants, Clifton, New Jersey, for Consolidated Edison Company of New York. Inhomogeneities in the crustal layers through which seismic waves travel account for the differences in arrival of P at similar distances. Location parameters were determined by WCC.

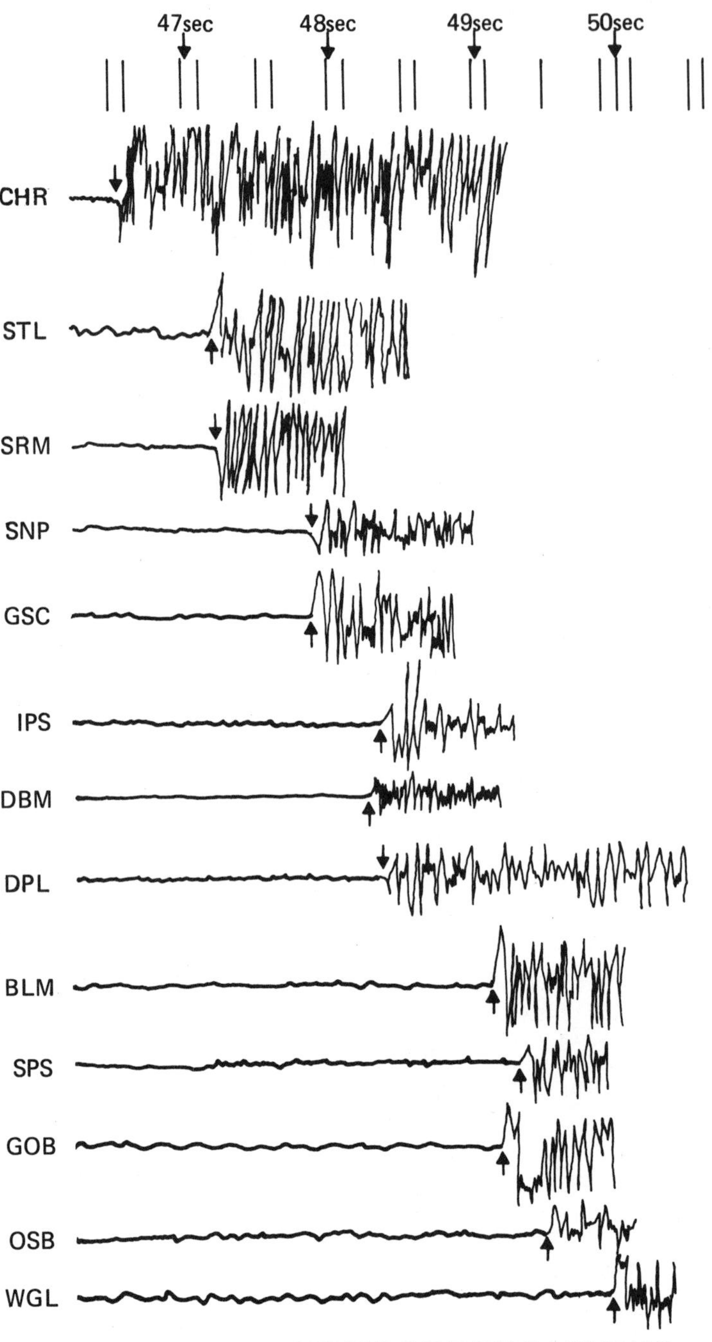

ARROWS INDICATE P ARRIVALS

These records were traced from the original

B 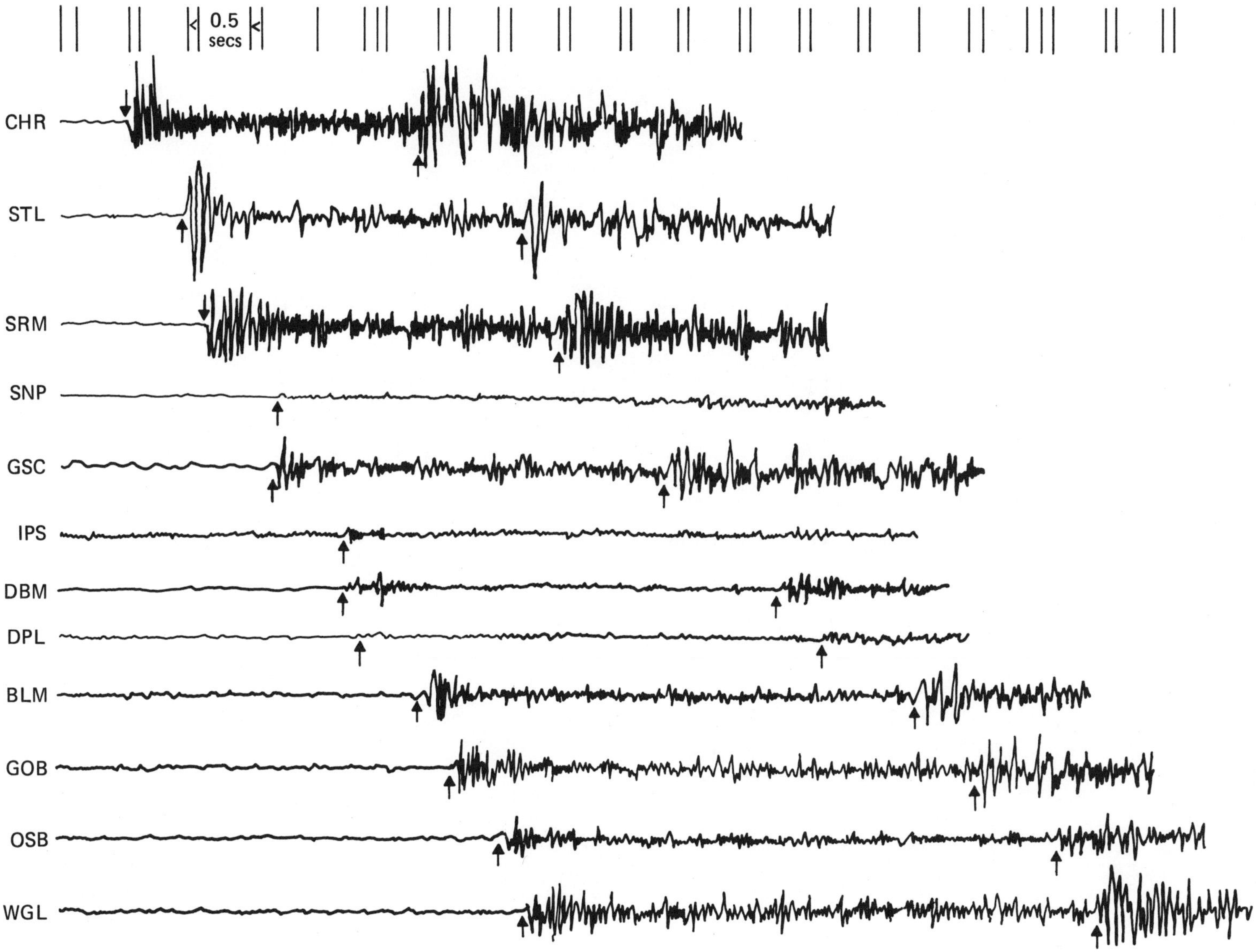
0.5 secs
CHR
STL
SRM
SNP
GSC
IPS
DBM
DPL
BLM
GOB
OSB
WGL
FIRST ARROW INDICATES P ARRIVAL
SECOND ARROW INDICATES S ARRIVAL
These records were traced from the original

Figure 2

	A	**B**
DISTANCE:	25 km	28 km
RECORDING STATION:	GOL	GOL
INSTRUMENTS:	SP Z, N, E	SP Z, N, E
LOCATION:	Loveland Pass, Colorado	Cabin Creek, Colorado
COORDINATES:	Undetermined	Undetermined
DATE:	5 January 1967	4 May 1967
TIME:	16 09 32.5 UTC (approximate)	13 29 36.0 UTC (approximate)
DEPTH:	0 km (estimated)	0 km (estimated)
MAGNITUDE:	Not computed	Not computed
COMMENTS:	Many of these events occurred during the 6 months of road construction on I-70 in this area.	The origin of this earthquake is unknown. This event, along with other events having this characteristic signature, occurred near a small reservoir completed in early 1967.

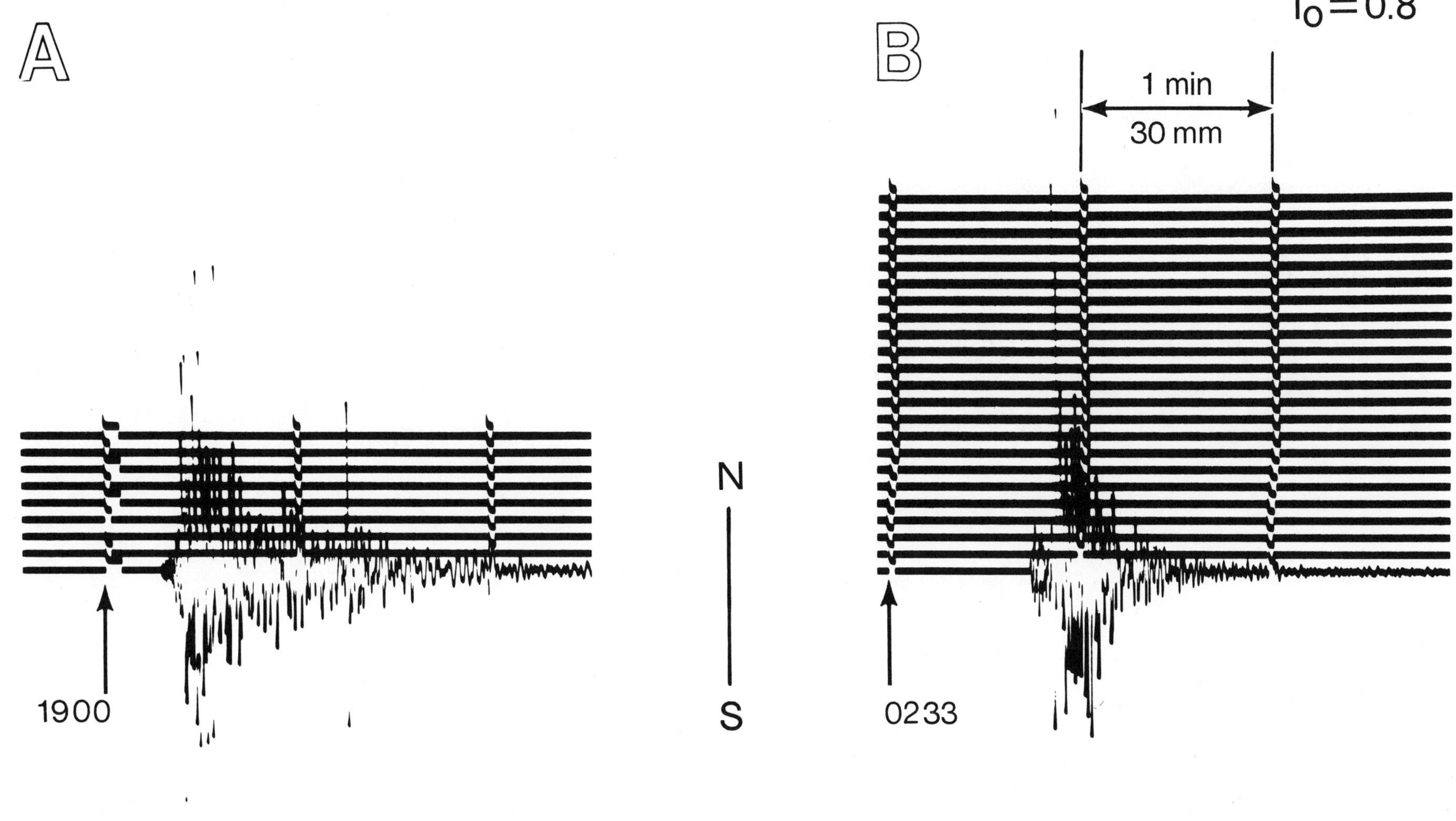

A
B
T$_O$=0.8
1 min
30 mm
N
S
1900
0233

Figure 4

DISTANCE:	40 km
RECORDING STATION:	GOL
INSTRUMENTS:	SP Z, N, E.
LOCATION:	Derby (northeast of Denver), Colorado
COORDINATES:	39.50 N, 105.00 W
DATE:	14 October 1966
TIME:	09 11 19.1 UTC (approximate)
DEPTH:	10 km (estimated)
MAGNITUDE:	2.4 M_L (GOL)

COMMENTS:

The total number of Derby earthquakes recorded at GOL between 1962 and 1967 was 1,604. The largest of these was 5.3 M_L on 9 August 1967, with maximum MM intensity VII. Fluid injection of wastes at the Rocky Mountain Arsenal disposal well were thought to have caused these earthquakes, which occurred in the fractured pre-Cambrian basement.

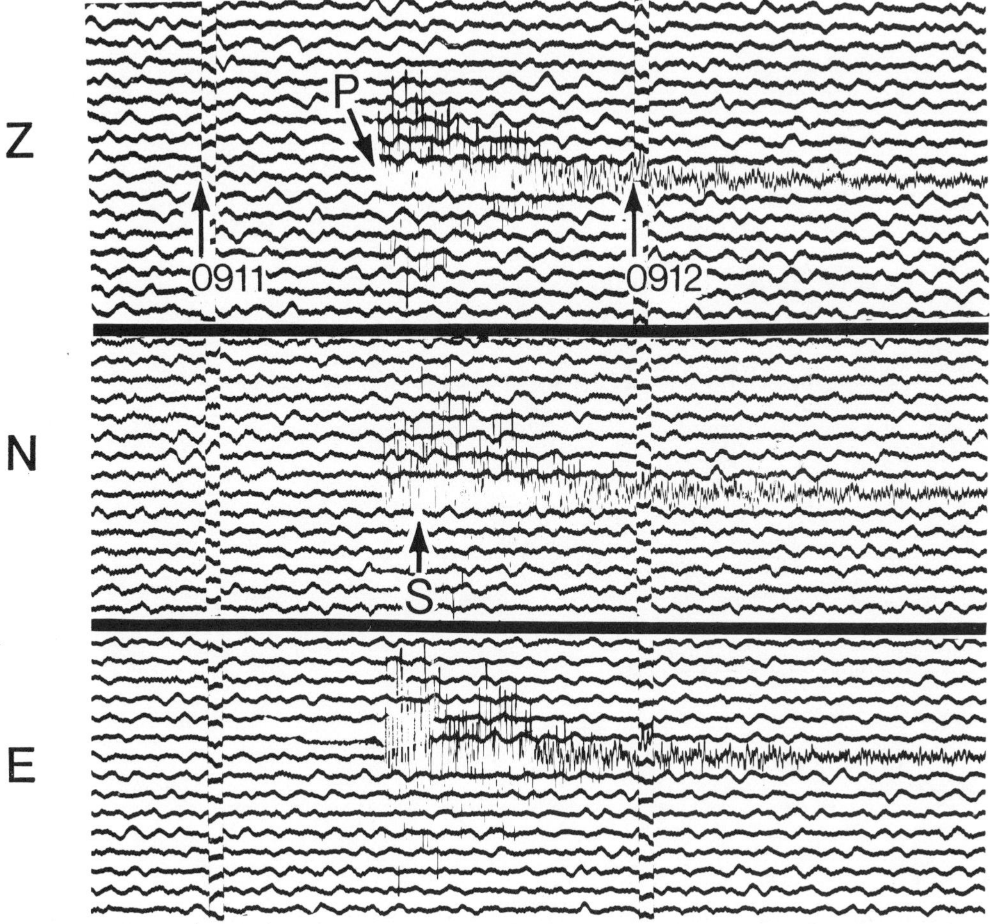

Z
N
E
P
S
0911
0912

Figure 5

DISTANCE: 75 km

RECORDING STATION: GOL

INSTRUMENTS: SP Z, N, E

LOCATION: Castle Rock, Colorado

COORDINATES: 39.38 N, 104.88 W (approximate)

DATE: 13 October 1966

TIME: 00 33 07.5 (approximate)

DEPTH: 0 km (estimated)

MAGNITUDE: Not computed

COMMENTS:

Extremely high frequencies occur throughout this near-regional earthquake or blast. There were no felt reports or sounds heard.

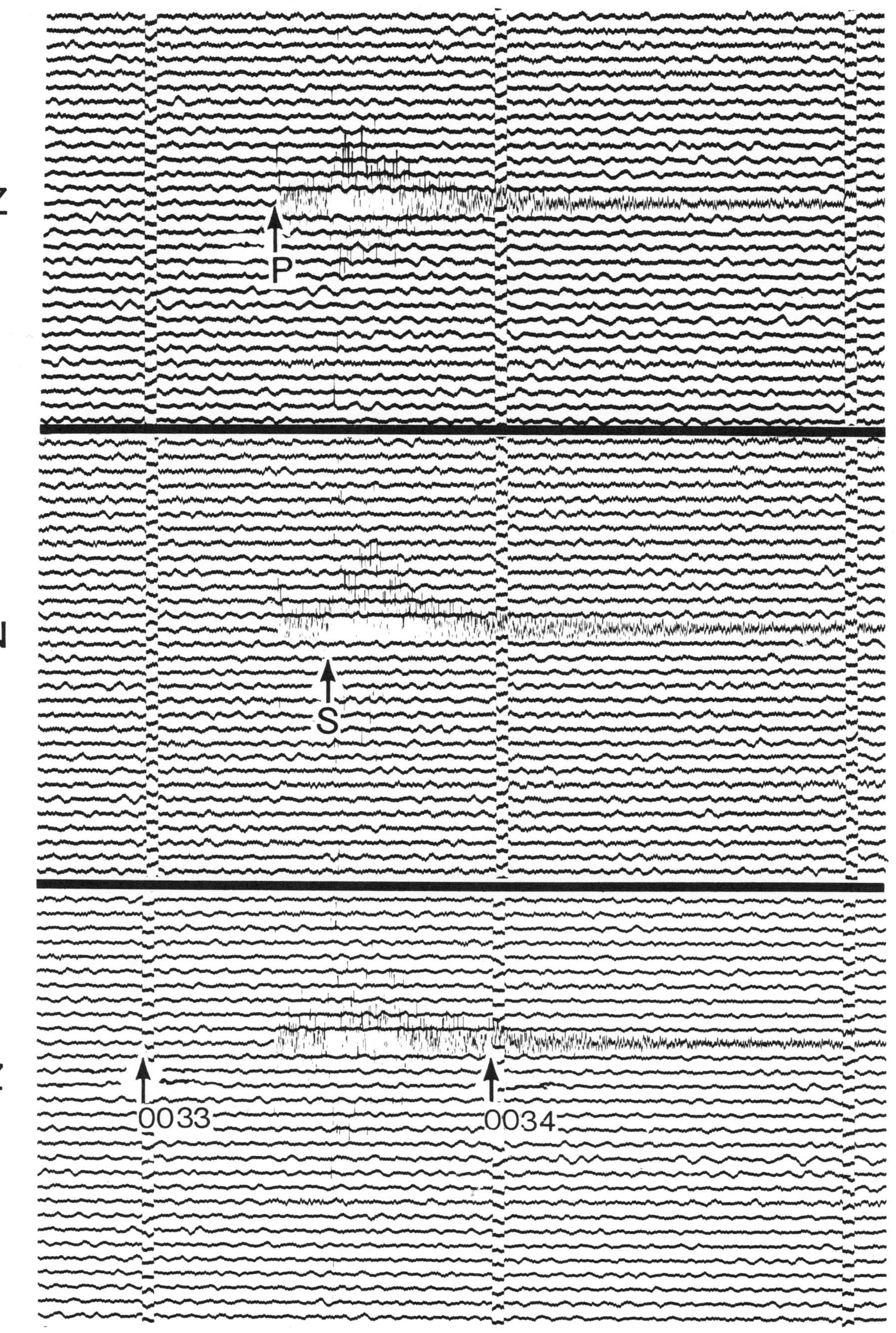

Z
N
Z
P
S
0033
0034

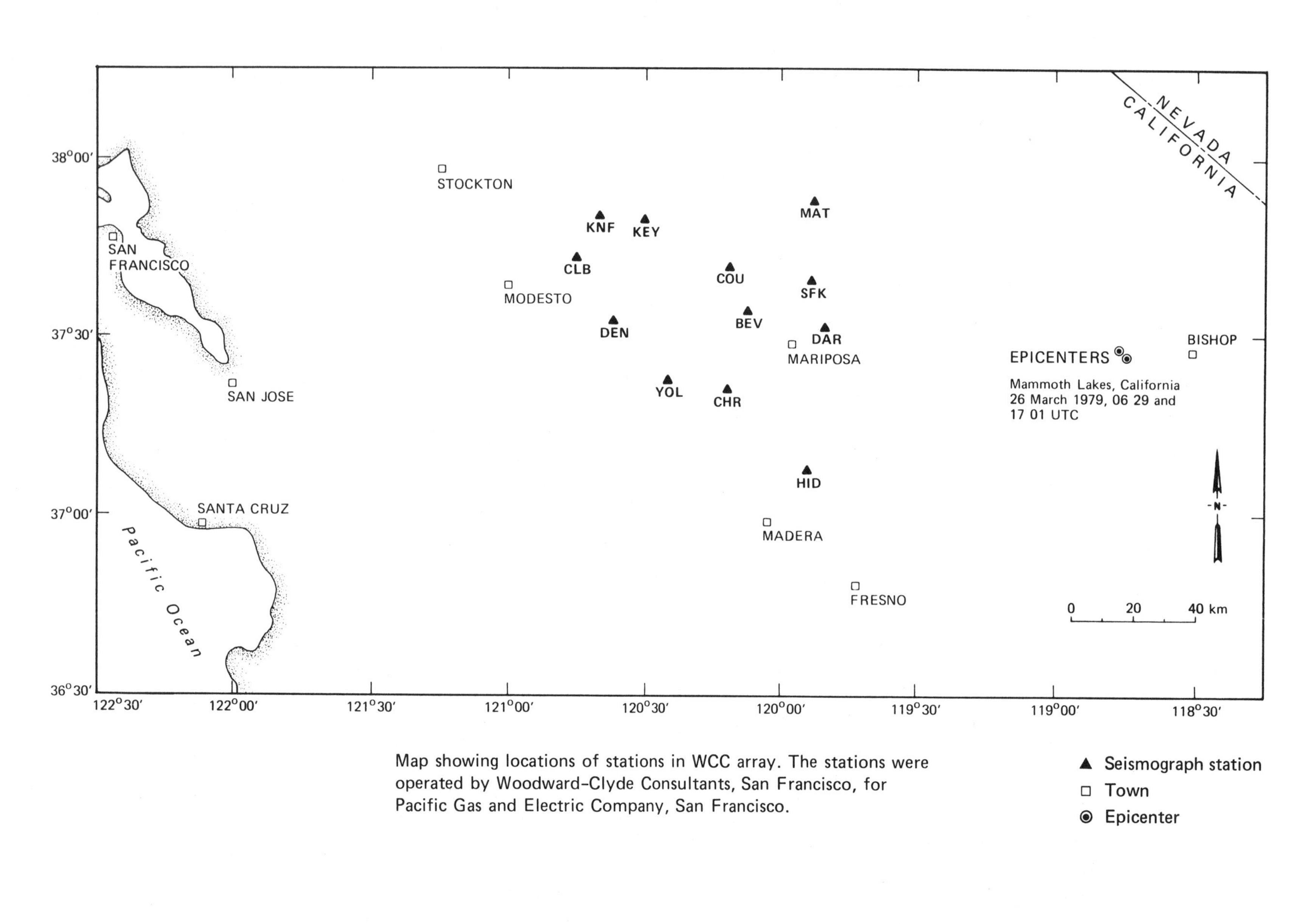

Map showing locations of stations in WCC array. The stations were operated by Woodward–Clyde Consultants, San Francisco, for Pacific Gas and Electric Company, San Francisco.

Figure 6 A and B

DISTANCE: 84 km to 190 km

RECORDING STATION: WCC array. See listing in Comments.

INSTRUMENTS: SP (1 Hz) Z, helicorder recordings

LOCATION: Mammoth Lakes, California

COORDINATES: 37.51 N, 118.69 W

DATE: 26 March 1979

TIME: 06 29 01.0 UTC

DEPTH: 5 km

MAGNITUDE: 3.2 M_L (BRK)

COMMENTS:

Station (see Map)	P Arrival Time (secs)	Distance (km)	Station (see Map)	P Arrival Time (secs)	Distance (km)
DAR	19.0	89	CHR	24.2	133
SFK	20.0	100	YOL	26.0	170
MAT	20.2	122	KEY	29.0	210
HID	21.0	125	DEN	30.0	190
BEV	23.0	130	KNF	31.0	200
COU	24.0	144	CLB	32.0	190

This suite of helicorder records was telemetered from a network operated by Woodward-Clyde Consultants, San Francisco, for Pacific Gas and Electric Company, San Francisco. P arrivals, listed above by increasing time and distance from the epicenter, indicate that crustal velocities are not uniform but vary from station to station. CHR has been omitted from the collection because of a differing time scale of recording. Location parameters were determined by WCC.

A

DAR
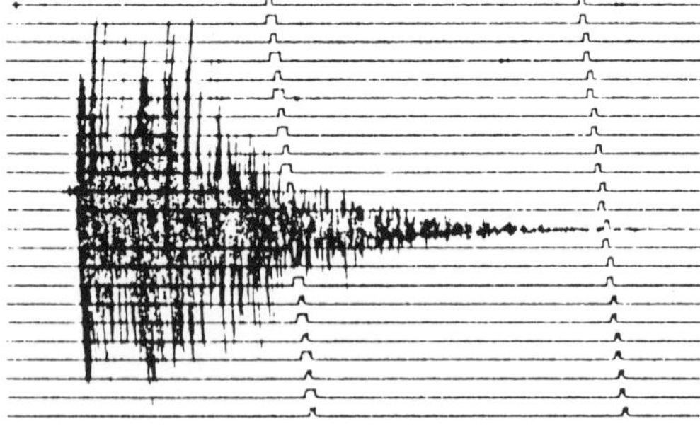
Z

SFK
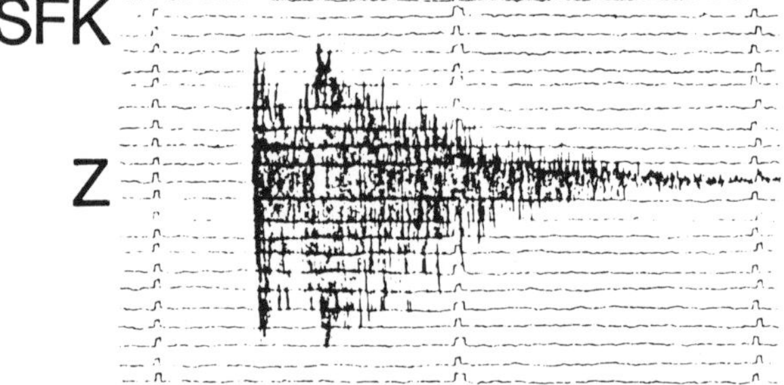
Z
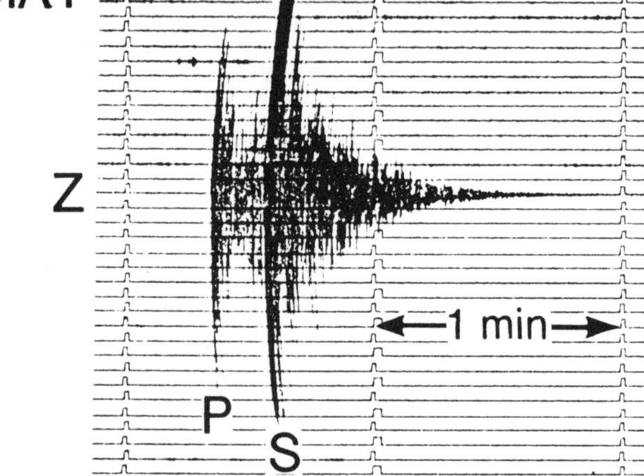
MAT
Z
1 min
P
S
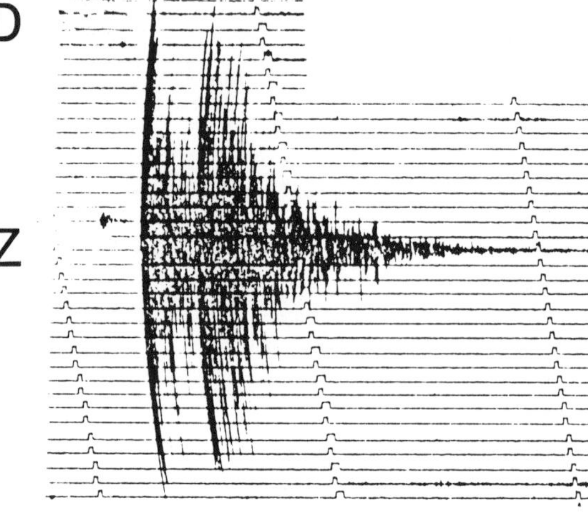
HID
Z
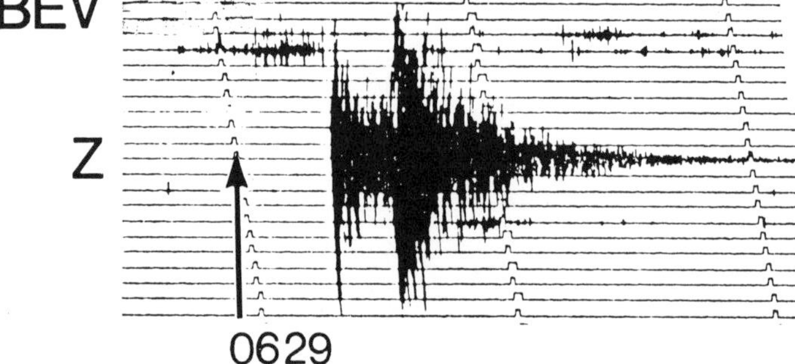
BEV
Z
0629
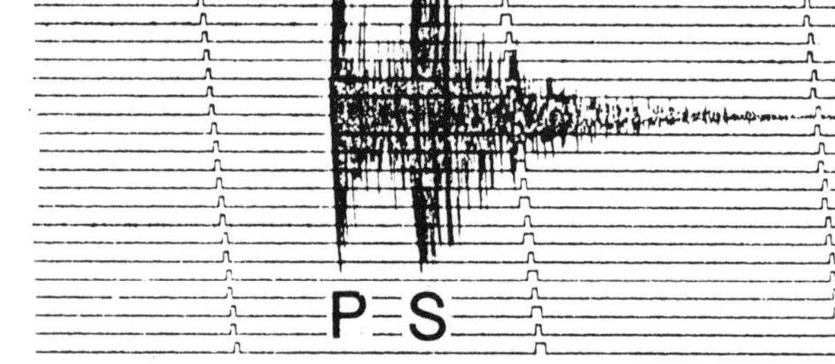
COU
Z
P S

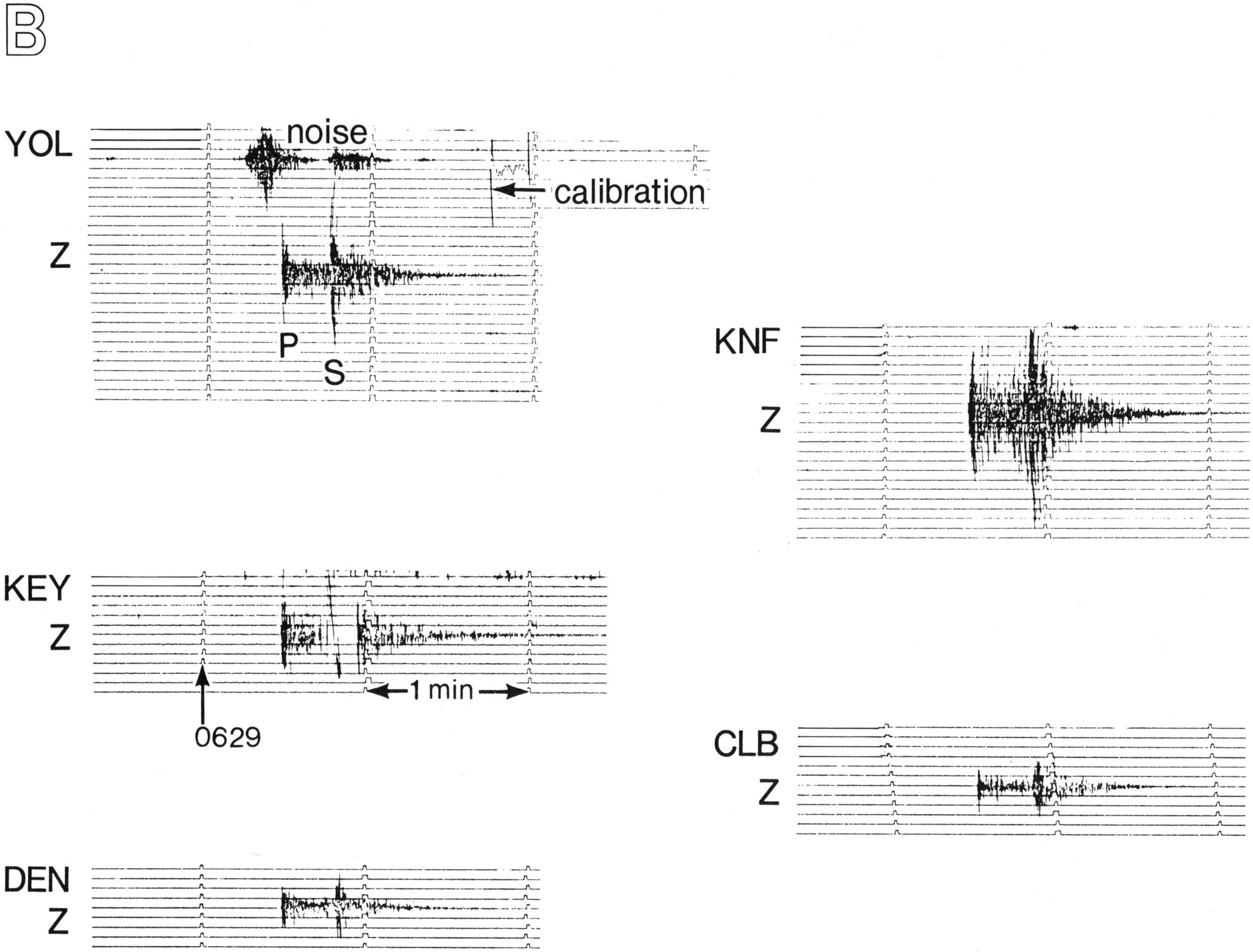

B
YOL
noise
calibration
Z
P
S
KNF
Z
KEY
Z
0629
1 min
CLB
Z
DEN
Z

Figure 7

DISTANCE:	140 km
RECORDING STATION:	GOL
INSTRUMENTS:	SP Z, N, E
LOCATION:	Aspen, Colorado
COORDINATES:	Undetermined
DATE:	19 December 1966
TIME:	21 52 38.2 UTC (approximate)
DEPTH:	Undetermined
MAGNITUDE:	3.3 M_L (GOL) (approximate)

COMMENTS:

This event was felt widely throughout western Colorado. There was no damage reported.

This is the closest distance at station GOL where Pn and Pg can be observed separately. This, however, may not be the case at other stations having different geologic characteristics.

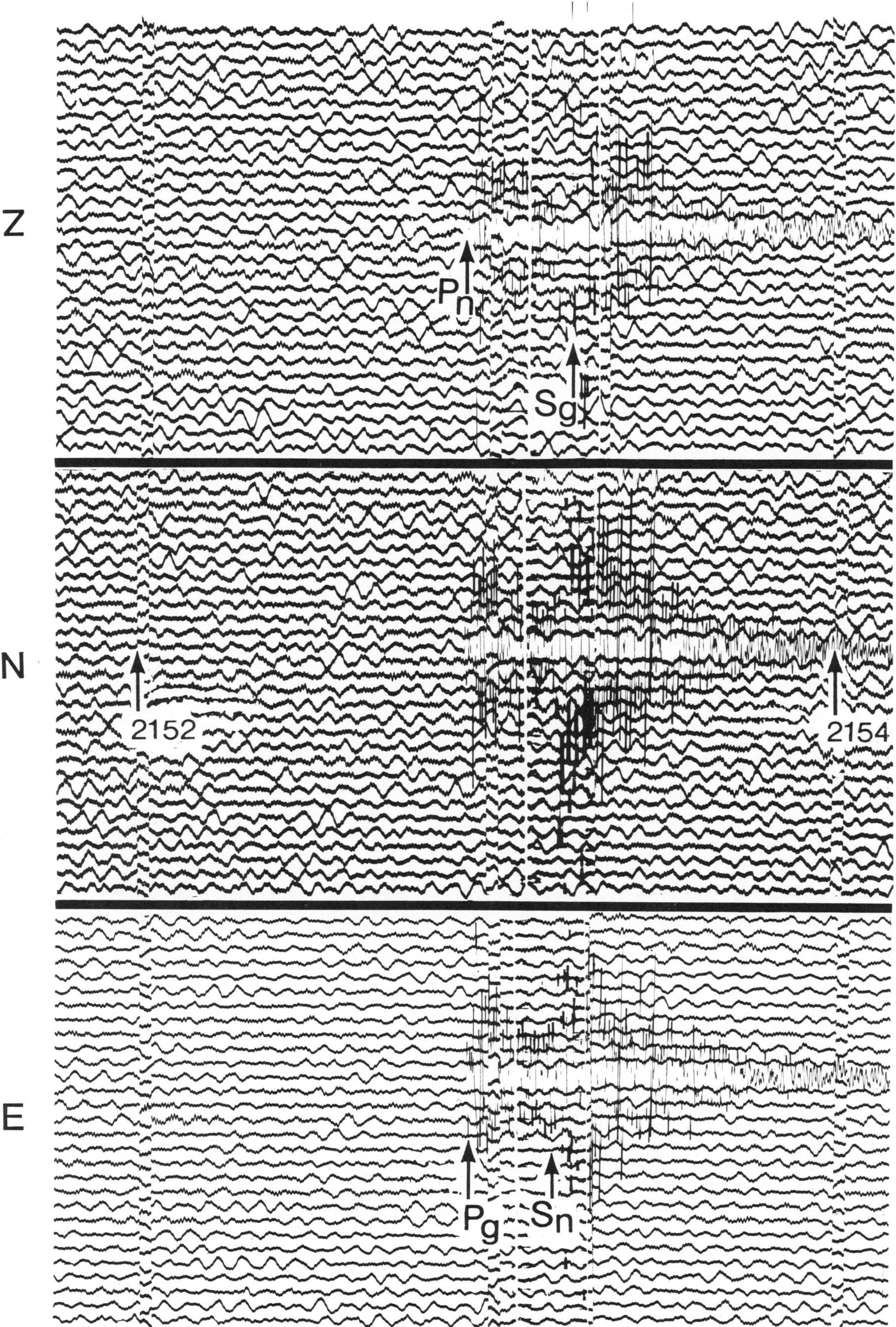

Z
Pn
Sg
N
2152
2154
E
Pg
Sn

Figure 8

DISTANCE: 200 km

RECORDING STATION: GOL

INSTRUMENTS: SP Z, N, E

LOCATION: Somerset, Colorado (Mt. Gunnison vicinity)

COORDINATES: Undetermined

DATE: 12 January 1967

TIME: 03 52 25.0 UTC (approximate)

DEPTH: Undetermined

MAGNITUDE: 3.8 M_L (GOL) (approximate)

COMMENTS:

A rockfall was reported from an inactive mine near Somerset at this time. One man was killed. This event was felt and heard in the area. Elkhead Mountains, at the same distance but at a NW azimuth, is the site of mine blasting, as well as earthquakes. Mine blasts, however, are much smaller than the record shown here.

There are almost equal amounts of P and S energy.

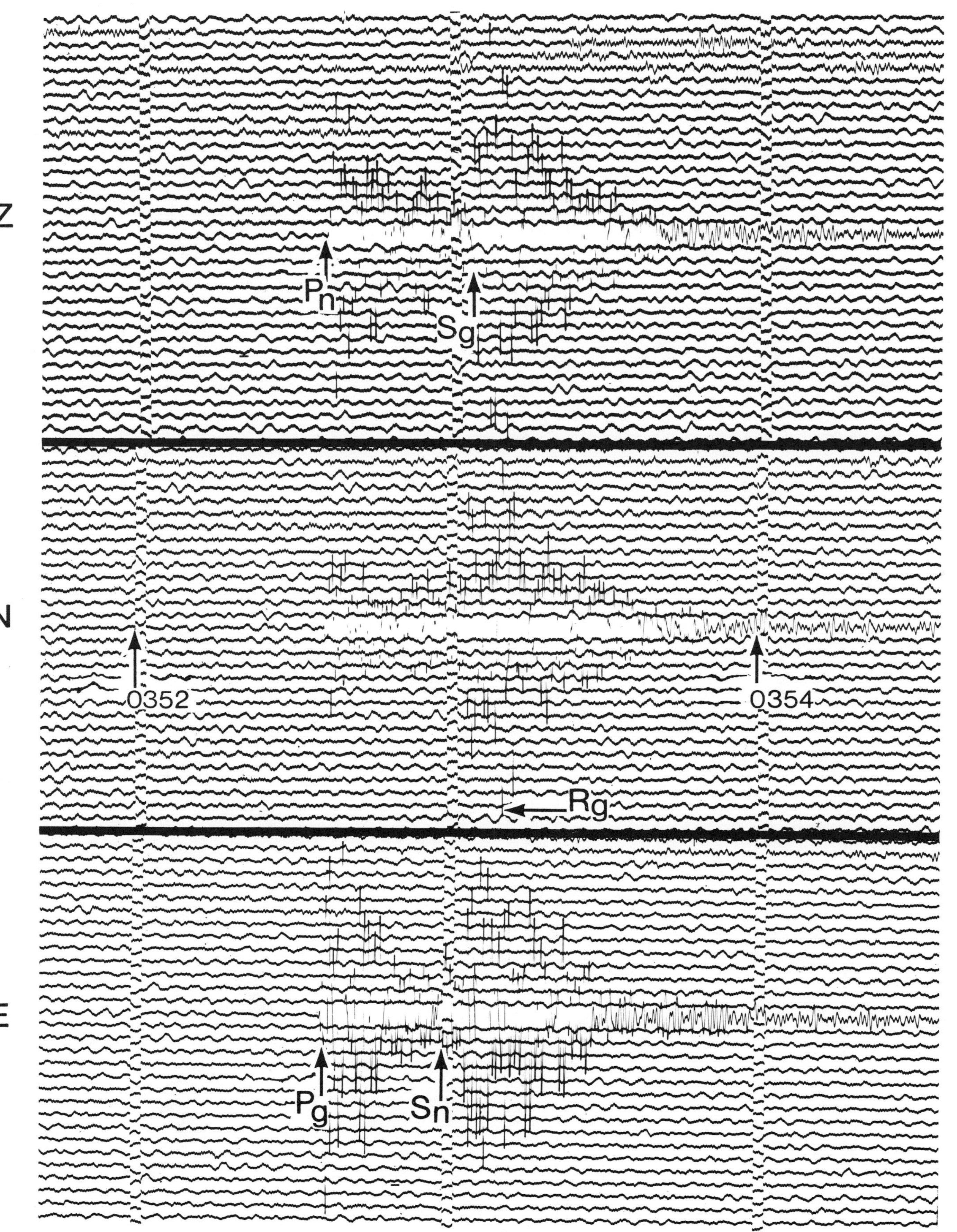

Z
N
E
Pn
Sg
O352
O354
Rg
Pg
Sn

Figure 9

DISTANCE: 220 km

RECORDING STATION: COL, College, Alaska (USGS)

INSTRUMENTS: SP Z, N, E

LOCATION: Alaska

COORDINATES: 63.20 N, 150.50 W

DATE: 13 February 1977

TIME: 23 45 37.9 UTC

DEPTH: 129 km

MAGNITUDE: 3.9 M_L (University of Alaska)

COMMENTS:

No clearcut Pn or Sn phases are seen on this regional earthquake because it was 129 km deep, an intermediate-depth earthquake.

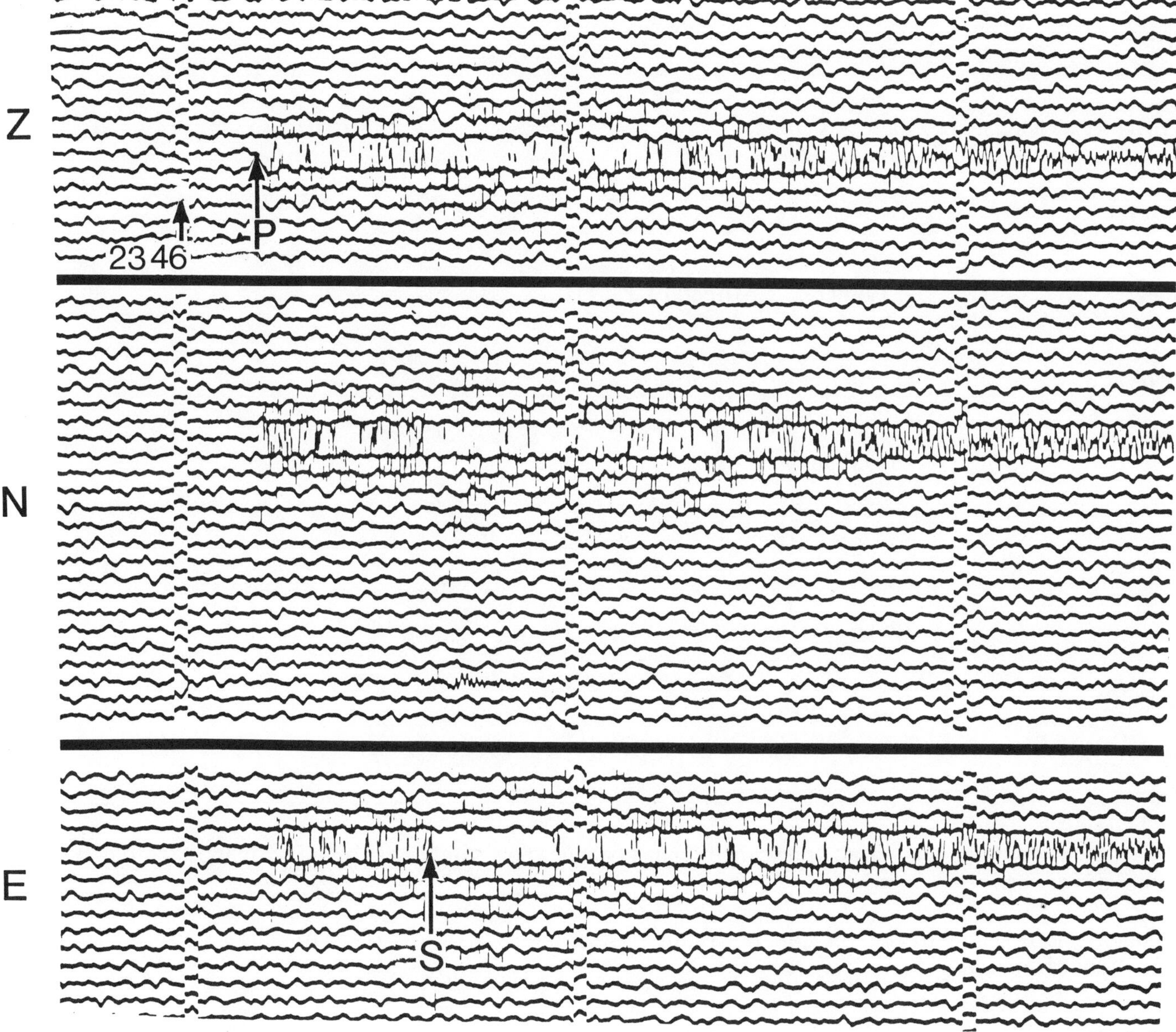

43

Figure 10

DISTANCE: 285 km

RECORDING STATION: GOL

INSTRUMENTS: SP Z, N, E

LOCATION: Silverton, Colorado

COORDINATES: 37.70 N, 107.80 W

DATE: 16 January 1967

TIME: 09 22 45.7 UTC

DEPTH: 33 km

MAGNITUDE: 4.1 M_L (USGS)

COMMENTS:

This widely felt earthquake, not an unusual event for southwestern Colorado, could have occurred in the silver mines.

The great amplitude of Pg on the horizontal components is typical of this region. The time of Sn is questionable.

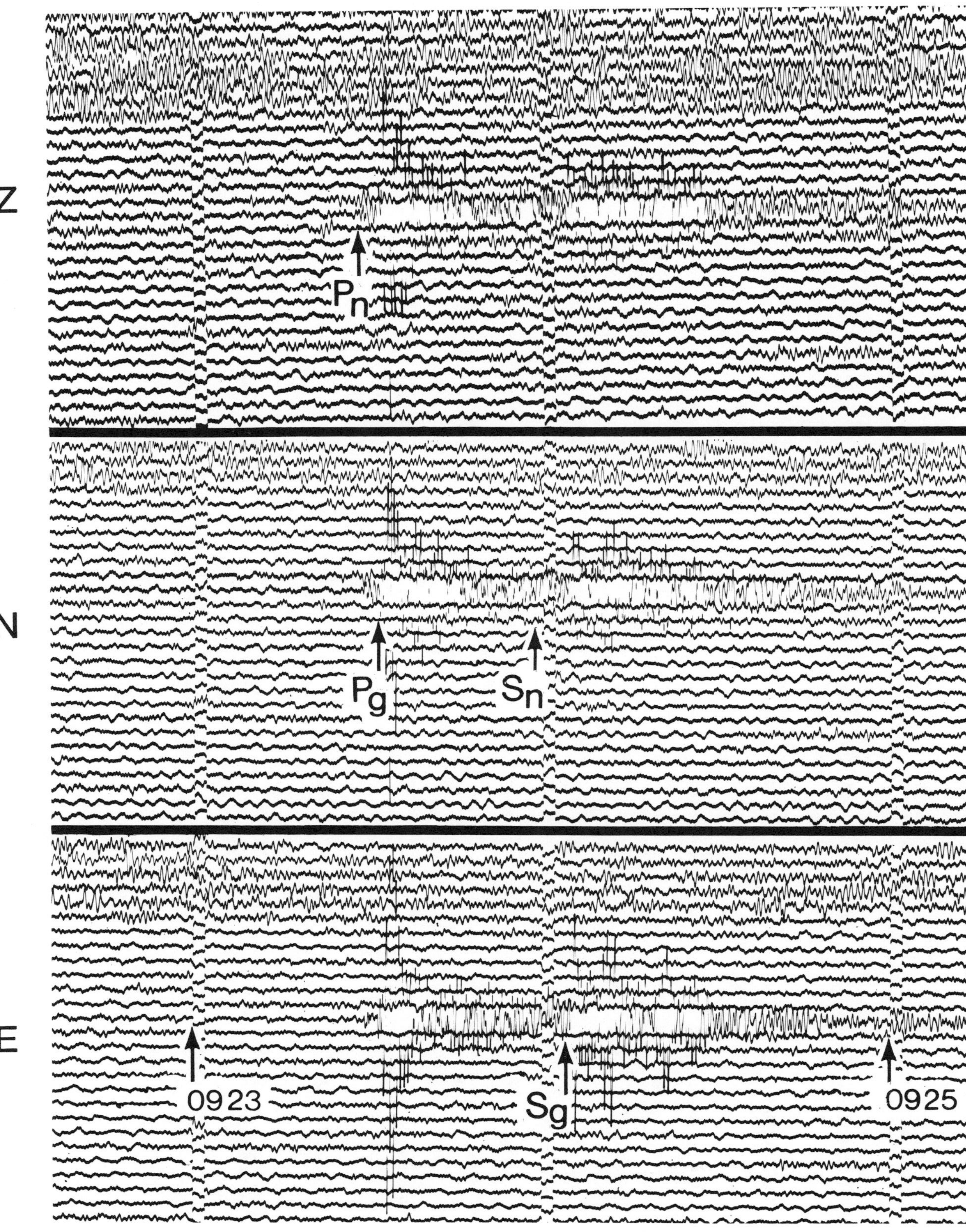

Z
N
E
Pn
Pg
Sn
Sg
09 23
0925

Figure 11

DISTANCE: 320 km

RECORDING STATION: JCT, Junction, Texas (USGS)

INSTRUMENTS: SP Z, N, E

LOCATION: West Texas

COORDINATES: 30.92 N, 103.11 W

DATE: 30 December 1974

TIME: 08 05 27.1 UTC

DEPTH: 5 km

MAGNITUDE: 3.7 M_L (USGS)

COMMENTS:

The highest amplitude waves in most regional earthquakes are the surface waves, which can be designated as L_g. In this earthquake, L_g is indistinguishable from S_g, seen in others at regional distances. Some earthquakes, such as the Colorado event at 200 km (Figure 8), have both; Sg is followed a few seconds later by a distinct L_g of even higher amplitude.

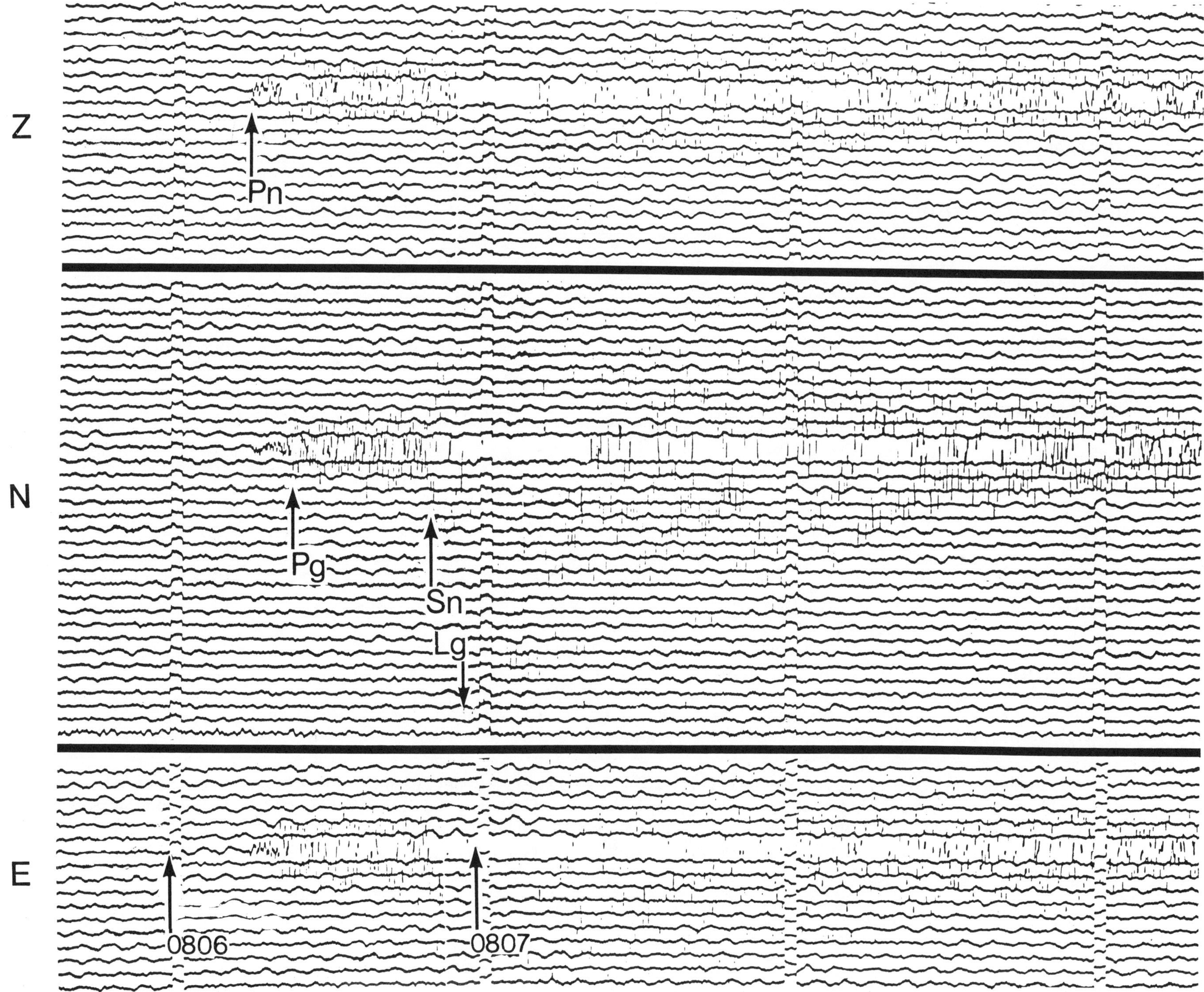

Z
Pn
N
Pg
Sn
Lg
E
0806
0807

Figure 12

	A	B
DISTANCE:	330 km	330 km
RECORDING STATION:	GOL	GOL
INSTRUMENTS:	SP Z, N, E	LP Z, N, E
LOCATION:	Dulce, New Mexico	Gasbuggy underground nuclear test, New Mexico
COORDINATES:	36.90 N, 107.00 W	36.70 N, 107.20 W
DATE:	6 January 1967	10 December 1967
TIME:	15 41 15.6 UTC	19 30 00.0 UTC
DEPTH:	33 km	1318 m
MAGNITUDE:	4.3 M_L (USGS)	4.6 M_L (GOL)

COMMENTS:

The high frequency waves seen after Pg on A records may represent reflections from several crustal layers. There were no long-period waves from the Dulce earthquake on the long-period records. The Gasbuggy underground nuclear test produced surface waves on the long-period components (B). These are sedimentary surface waves having a velocity of approximately 2.5 km/sec.

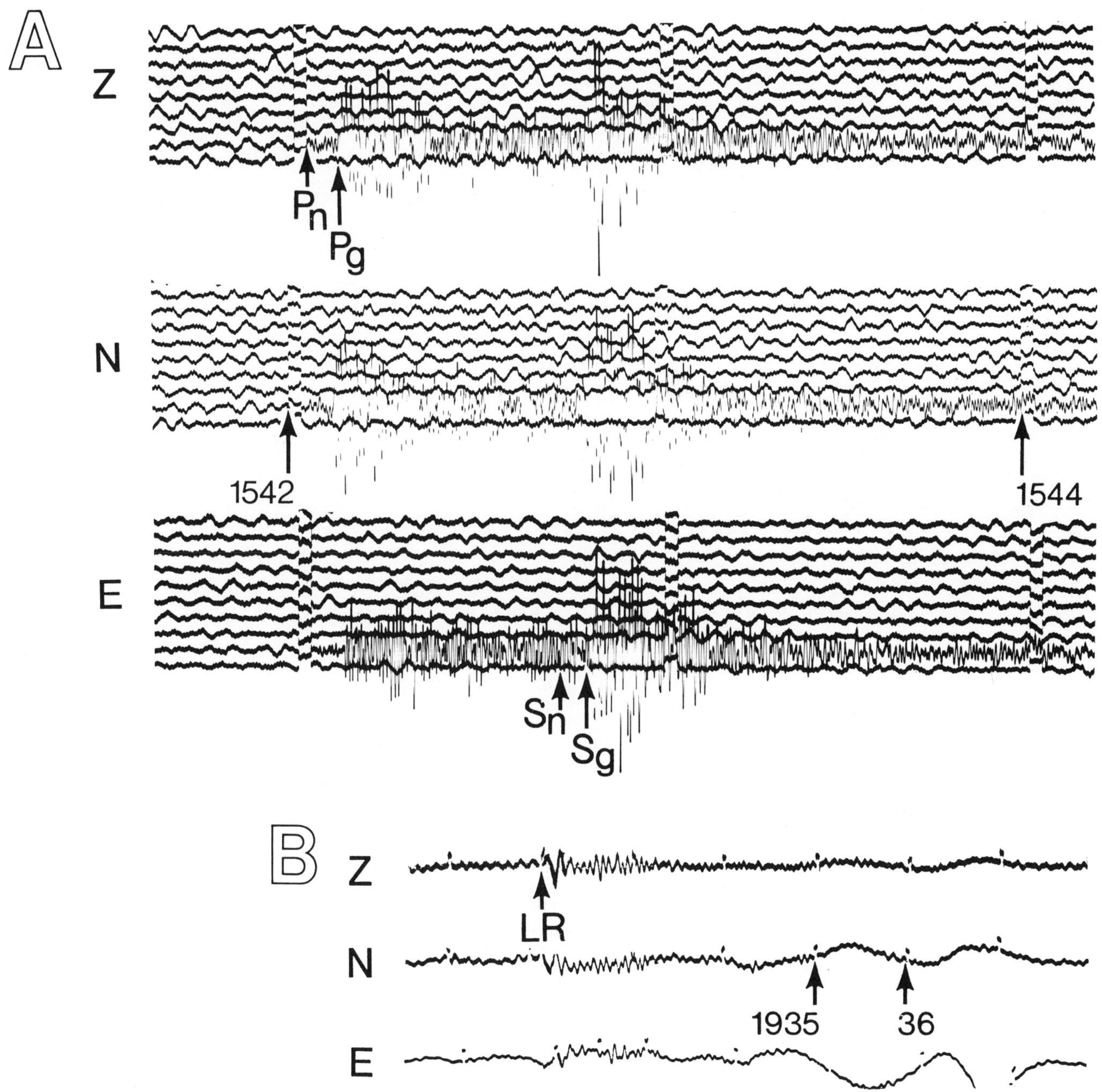

A
Z
Pn
Pg
N
1542
1544
E
Sn
Sg
B
Z
LR
N
1935
36
E

Figure 13

	A	B	C
DISTANCE:	366 km	400 km	610 km
RECORDING STATION:	TNP, Tonopah, Nevada	KPK, Kanaka Peak, California	BMN, Battle Mt, Nevada
INSTRUMENTS:	SP (1 Hz) Z, USGS helicorder records	SP (1 Hz) Z, USGS helicorder records	SP (1 Hz) Z, USGS helicorder records
LOCATION:	Bradley, San Miguel vicinity, California	Bradley, San Miguel vicinity, California	Bradley, San Miguel vicinity, California
COORDINATES:	35.96 N, 120.49 W	35.96 N, 120.49 W	35.96N, 120.49 W
DATE:	29 November 1977	29 November 1977	29 November 1977
TIME:	16 42 02.4 UTC	16 42 02.4 UTC	16 42 02.4 UTC
DEPTH:	11 km	11 km	11 km
MAGNITUDE:	3.7 M_L (BRK)	3.7 M_L (BRK)	3.7 M_L (BRK)

COMMENTS:

This earthquake has been recorded at three stations at slightly increasing distances, with different azimuths from the epicenter. Pn is emergent, whereas Pg is more impulsive on each record.

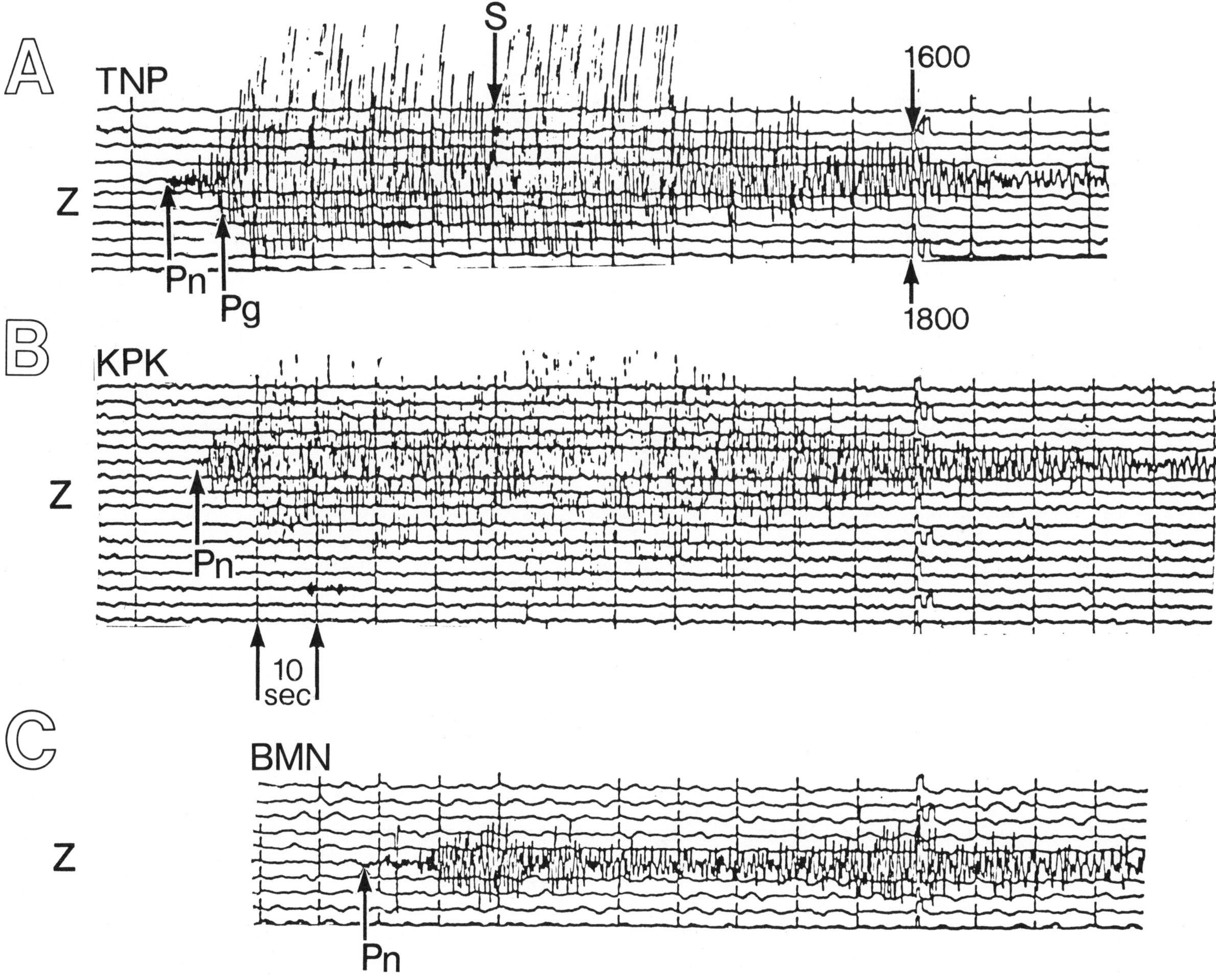

A
TNP
S
1600
Z
Pn
Pg
1800
B
KPK
Z
Pn
10 sec
C
BMN
Z
Pn

Figure 14

DISTANCE:	455 km
RECORDING STATION:	JCT Junction, Texas (USGS)
INSTRUMENTS:	SP Z, N, E
LOCATION:	New Mexico
COORDINATES:	32.31 N, 104.14 W
DATE:	28 November 1974
TIME:	03 35 20.5 UTC
DEPTH:	5 km
MAGNITUDE:	3.9 M_L (USGS)

COMMENTS:

Pn and Pg are well developed phases. Lg begins where the trace disappears because of the high-amplitude surface waves. Sn is more prominent on the horizontal components.

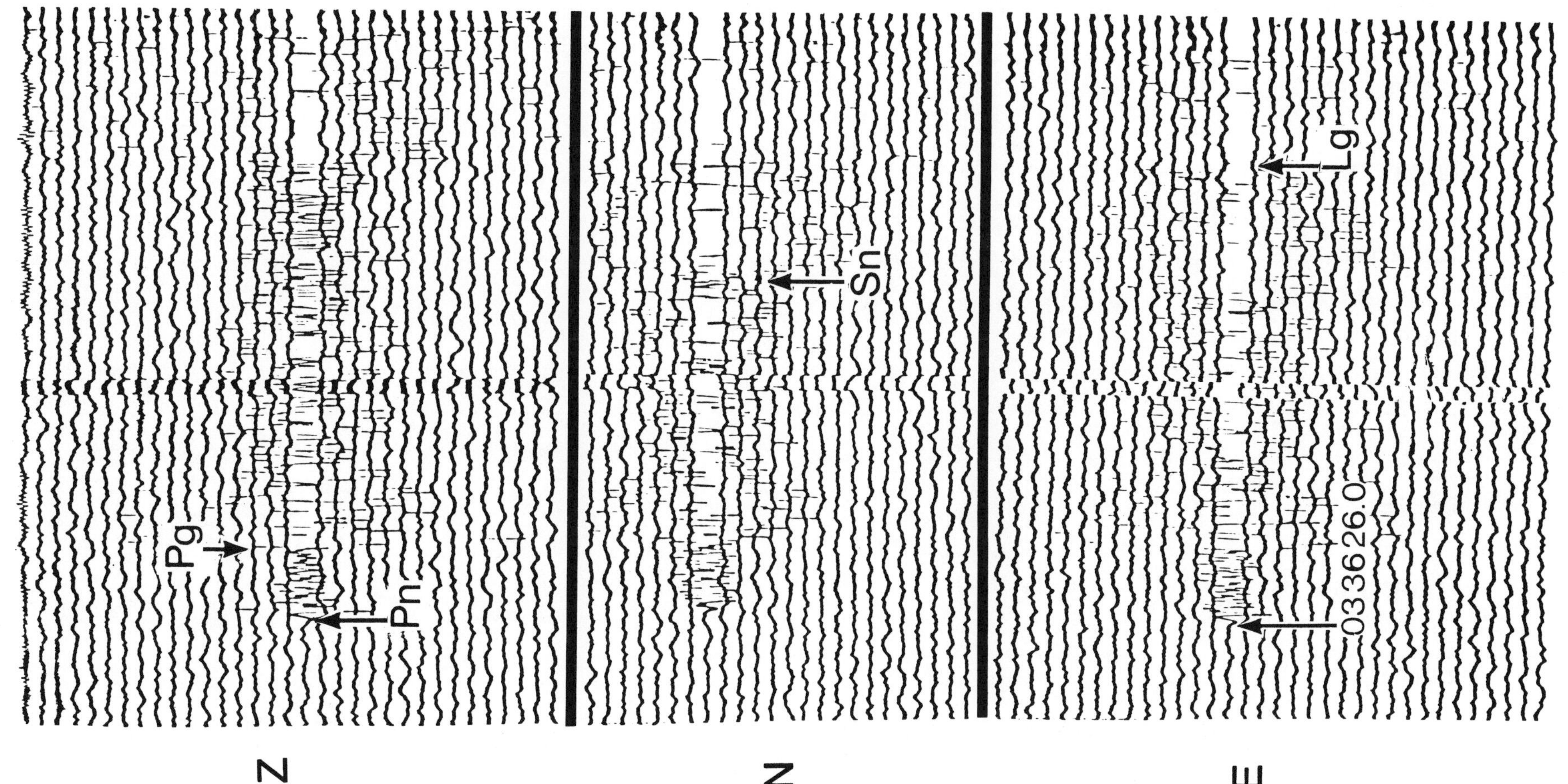

Pg
Pn
Sn
Lg
033626.0
Z
N
E

Figure 15

DISTANCE:	456 km
RECORDING STATION:	GOL
INSTRUMENTS:	SP Z, N, E
LOCATION:	Monument Butte, Wyoming
COORDINATES:	42.00 N, 110.2 W
DATE:	10 March 1967
TIME:	02 20 35.4 UTC
DEPTH:	33 km
MAGNITUDE:	3.6 M_L (USGS)

COMMENTS:

This is the nearest distance at station GOL at which P* can be seen clearly. Other stations having different geologic conditions may see this phase at closer distances. When P* is seen, there is often also an S*.

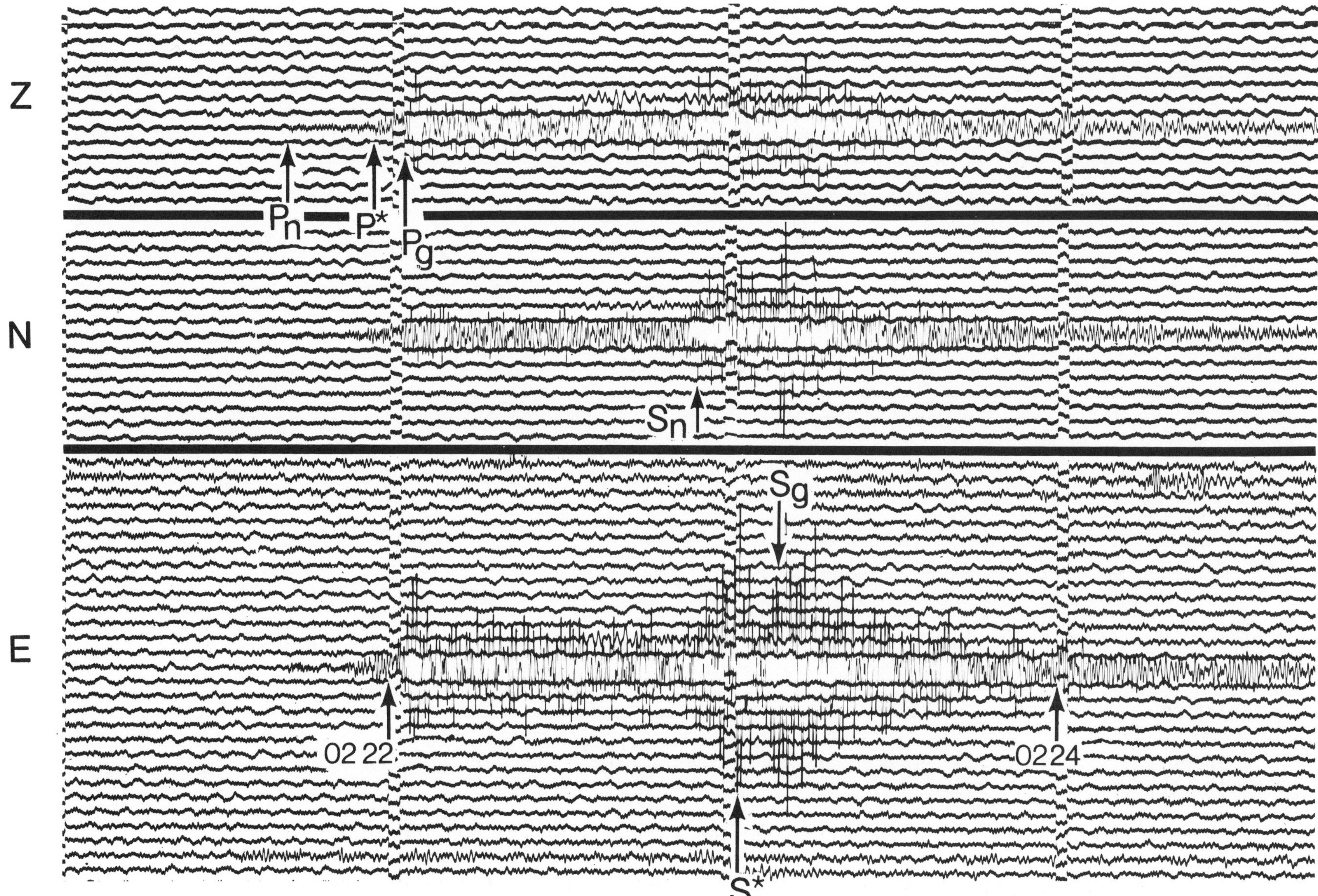

Z
N
E
P_n
P*
P_g
S_n
S_g
S*
02 22
02 24

Figure 16

DISTANCE: 5.3° (600 km)

RECORDING STATION: GOL

INSTRUMENTS: LP Z, N, E

LOCATION: Richfield, Utah

COORDINATES: 38.50 N, 112.10 W

DATE: 4 October 1967

TIME: 10 20 14.0 UTC

DEPTH: 18 km

MAGNITUDE: 5.2 M_L (USGS), 4 3/4 to 5 M_L (PAS), 5.4 to 5.7 M_L (BRK), 5.5 M_L (GOL)

COMMENTS:

There was slight damage reported in the Marysvale area. The earthquake was felt strongly throughout southern Utah.

The high frequencies following P and S on these records are of considerable research interest for those studying crustal structure. The unusual surface wave dispersion for a regional earthquake may be caused by the complicated structure of the Rocky Mountains. The short-period components were off scale during the earthquake.

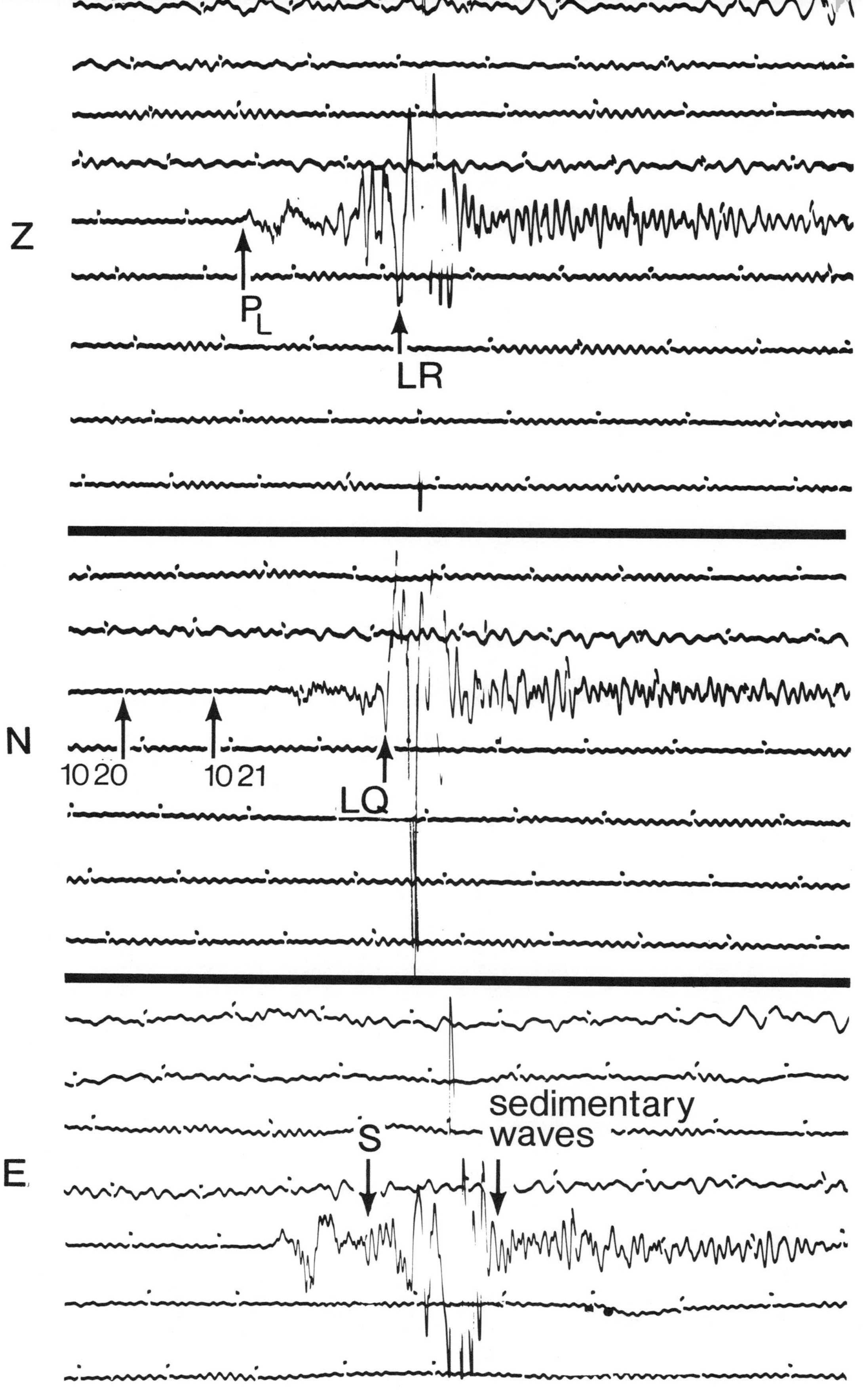
Z
P
L
LR
N
10 20
10 21
LQ
sedimentary
waves
S
E

Figure 17

DISTANCE:	5.9°
RECORDING STATION:	GOL
INSTRUMENTS:	SP Z, N, E
LOCATION:	Eastern Idaho
COORDINATES:	42.08 N, 112.45 W
DATE:	22 September 1975
TIME:	10 42 36.2 UTC
DEPTH:	3 km
MAGNITUDE:	4.2 M_L (USGS), 3.6 M_L (SLC)

COMMENTS:

Felt (MM intensity IV) at Holbrook and Portage.

The emergent Pn is followed by high-frequency waves in both Pn and Pg, up to and including Sn, which can be seen best on the N-S component.

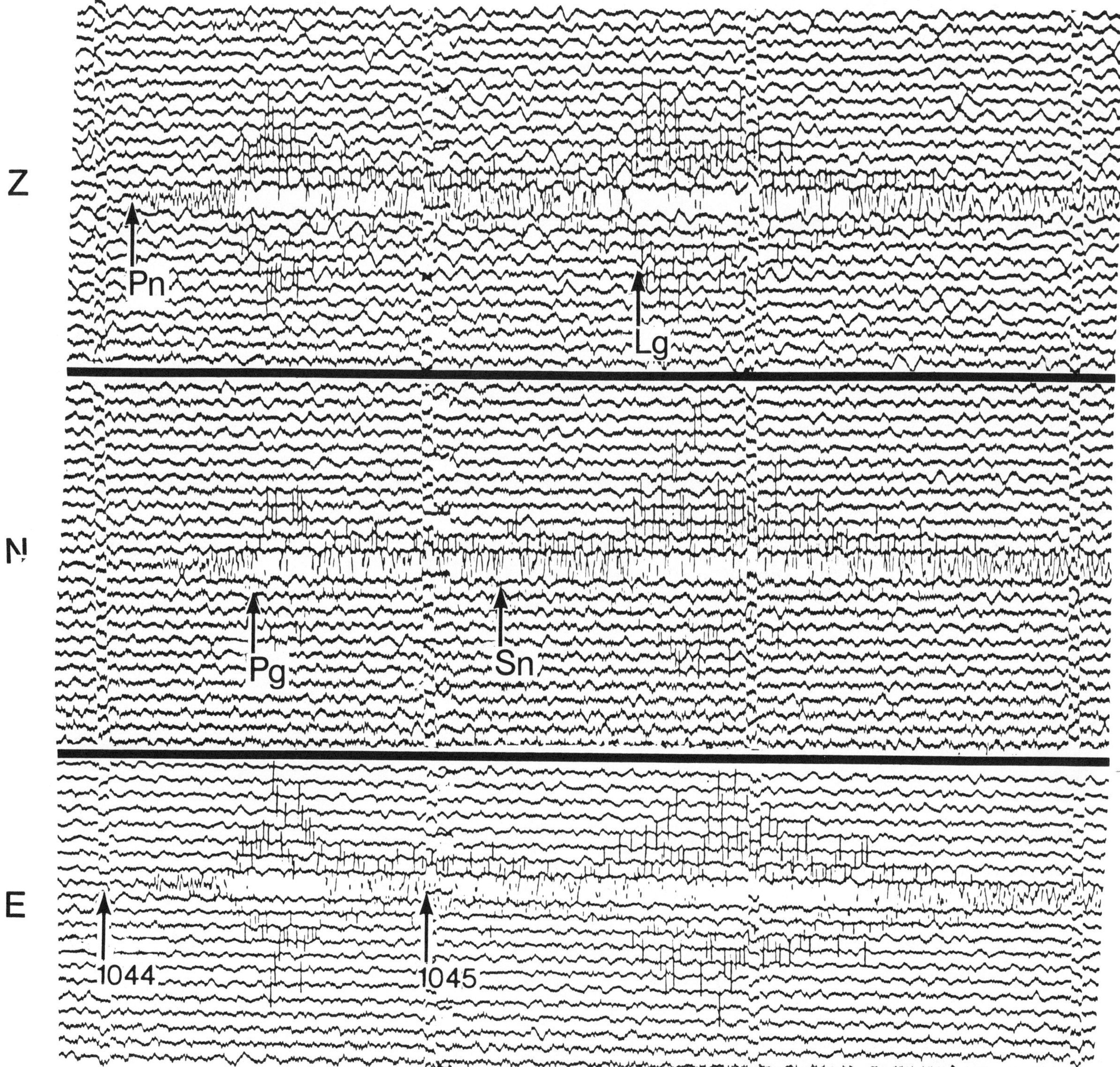

Z
Pn
Lg
N
Pg
Sn
E
1044
1045

Figure 18

DISTANCE: 6°

RECORDING STATION: GOL

INSTRUMENTS: SP Z, N, E

LOCATION: Winner, South Dakota

COORDINATES: 43.70 N, 99.40 W

DATE: 23 November 1967

TIME: 06 23 38.6 UTC

DEPTH: 16 km

MAGNITUDE: 4.4 M_L (USGS)

COMMENTS:

This earthquake was widely felt in southern South Dakota and northern Nebraska.

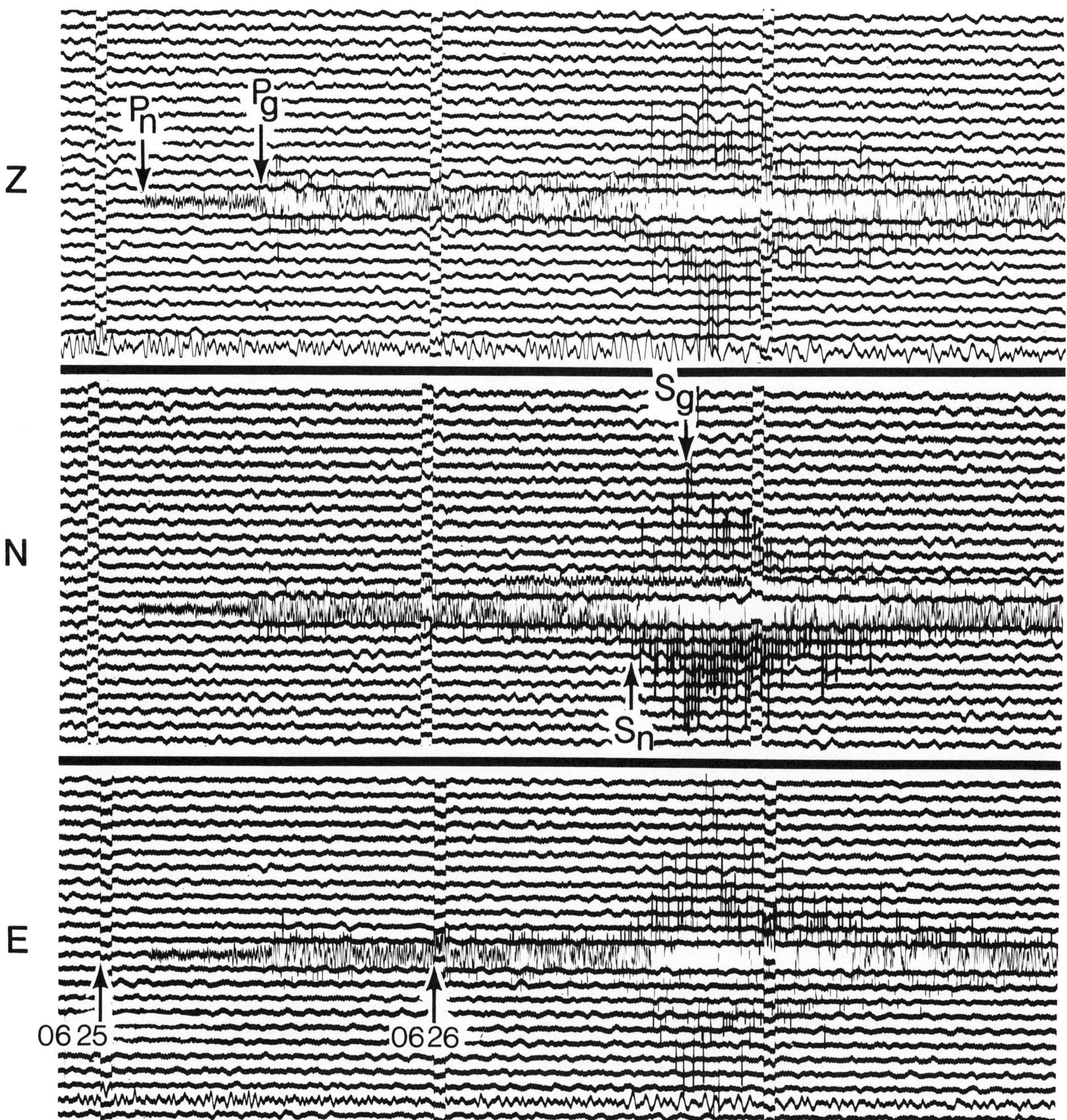

Z
N
E
Pn
Pg
Sg
Sn
06 25
06 26

Figure 19

DISTANCE: 8°

RECORDING STATION: MGS (USGS, South Carolina network)

INSTRUMENTS: SP Z

LOCATION: North Atlantic Ocean (500 km SW of Bermuda)

COORDINATES: 29.80 N, 67.40 W

DATE: 24 March 1978

TIME: 00 42 36.3 UTC

DEPTH: 20 km

MAGNITUDE: 6.1 M_b, 5.8 M_s (USGS), 6.1 M_s (BRK), 6.0 M_s (PAS)

COMMENTS:

This earthquake was reported felt on Bermuda, at Boston, Massachusetts, and in parts of North Carolina, with P waves reported from 272 stations around the world. The epicenter was not on the seismic Mid-Atlantic Ridge, but a considerable distance west of it, making this an intraplate earthquake. Both the size and location of this earthquake are unusual.

The T phase illustrated here is a short-period wave train traveling as a sound wave of low velocity in the sea. It is propagated from the shore inland with the velocity of seismic waves. These are guided or trapped waves, channeled in the crust, and can be considered similar to L_g or P_L, which are also guided waves in the upper crustal layers of the earth. The T wave arrived, in this case, about nine minutes after the P wave.

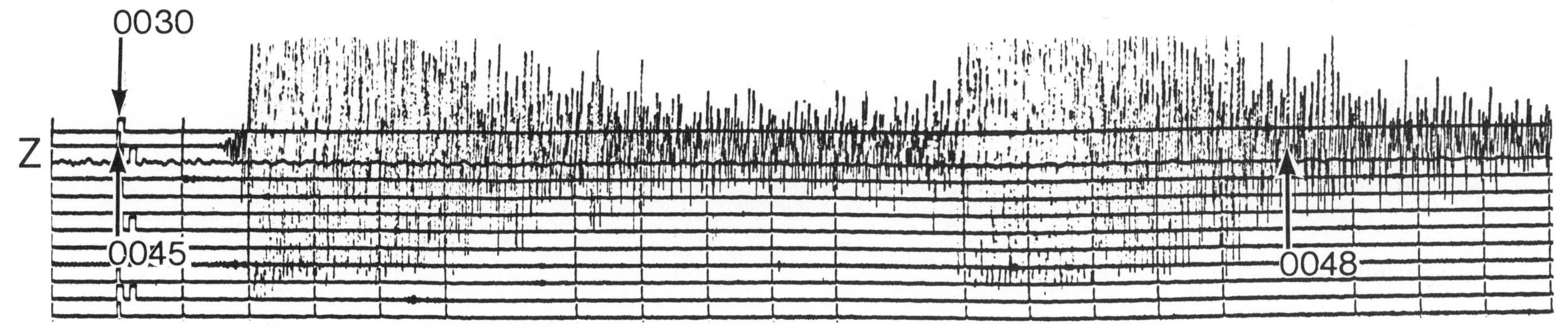

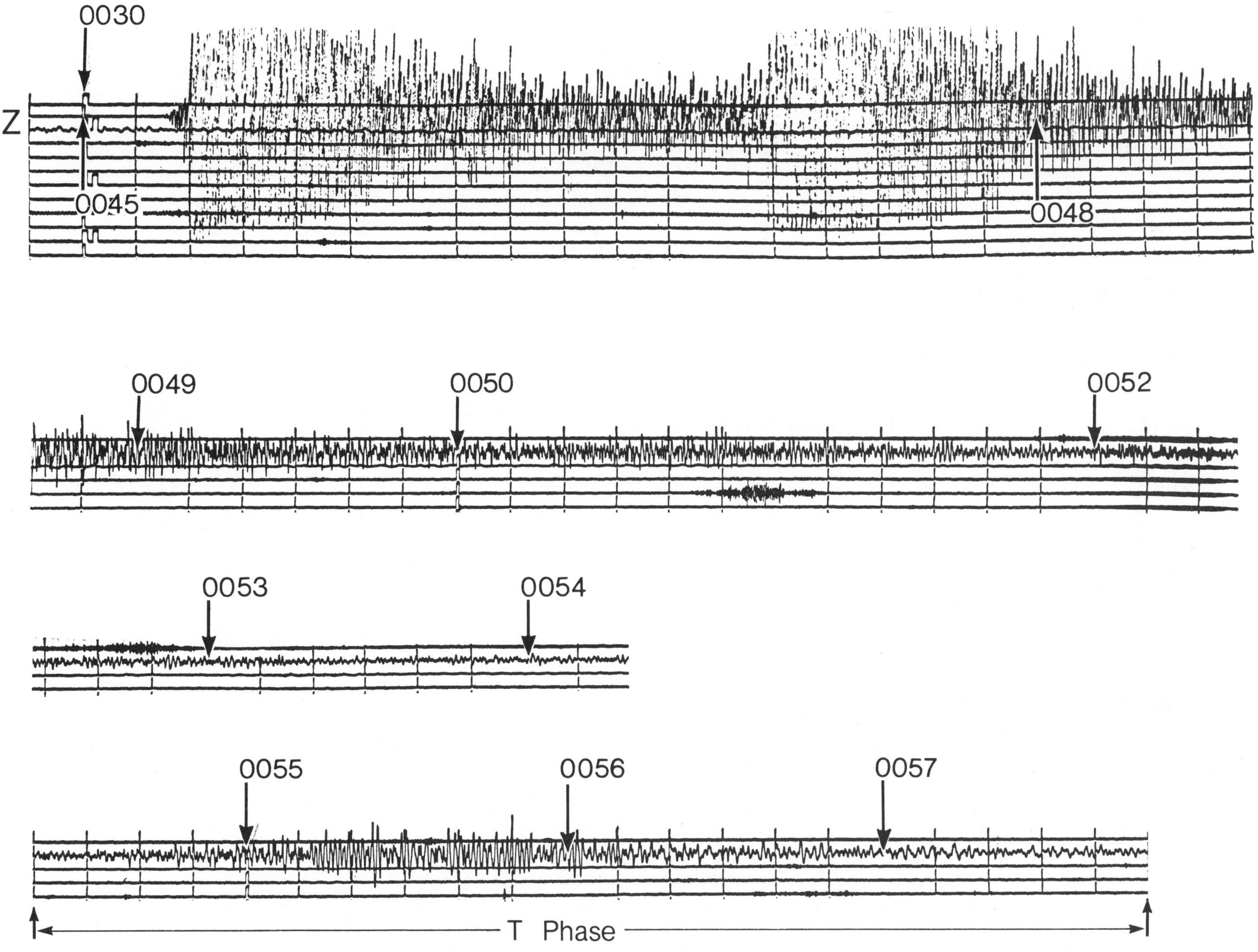

0030
Z
0045
0048
0049
0050
0052
0053
0054
0055
0056
0057
T Phase

Figure 20

DISTANCE:	9°
RECORDING STATION:	GOL
INSTRUMENTS:	SP Z, N, E
LOCATION:	Yucca Flats, Nevada (Nuclear Test Site)
COORDINATES:	37.00 N, 115.80 W
DATE:	19 January 1967
TIME:	16 45 00.0 UTC
DEPTH:	0 km
MAGNITUDE:	4 3/4 to 5 M_b (GOL)

COMMENTS:

This underground nuclear explosion produced a weak S wave, strong Pn, P*, and Pg, but few long-period surface waves at this magnitude. Compare these records with those of Figure 21, which show two other nuclear explosions of larger magnitude.

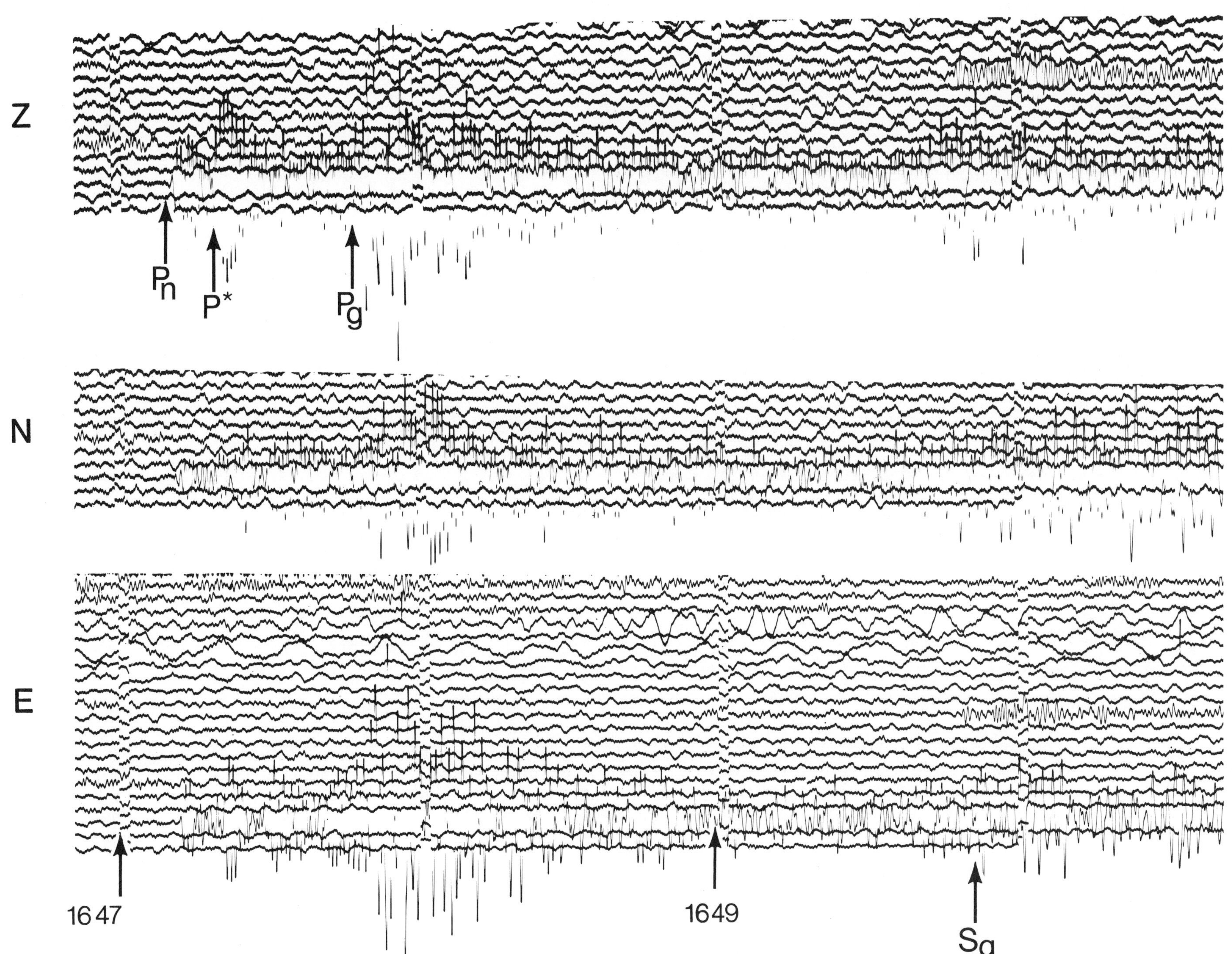

Z
N
E
P_n
P*
P_g
16 47
16 49
S_g
65

Figure 21

	A	B
DISTANCE:	9°	9°
RECORDING STATION:	GOL	GOL
INSTRUMENTS:	LP Z, N, E	LP Z, N, E
LOCATION:	Yucca Flats, Nevada (Nuclear Test Site)	Yucca Flats, Nevada (Nuclear Test Site)
COORDINATES:	37.00 N, 115.80 W	37.00 N, 115.80 W
DATE:	23 May 1967	20 May 1967
TIME:	14 00 00.0 UTC	15 00 00.0 UTC
DEPTH:	0 km	0 km
MAGNITUDE:	5.8 M_S (GOL)	6.1 M_S (GOL)

COMMENTS:

There is a slight difference in surface-wave configuration, despite the identical location of these underground nuclear explosions. The short-period P waves were off scale for both events. The collapse of the cavity occurred 14 minutes after the origin time of the explosion. This can be seen on all the long-period components in B. Collapse time can follow the explosion by 24 hours or longer, in which case the surface wave can be misinterpreted as an earthquake at the same distance.

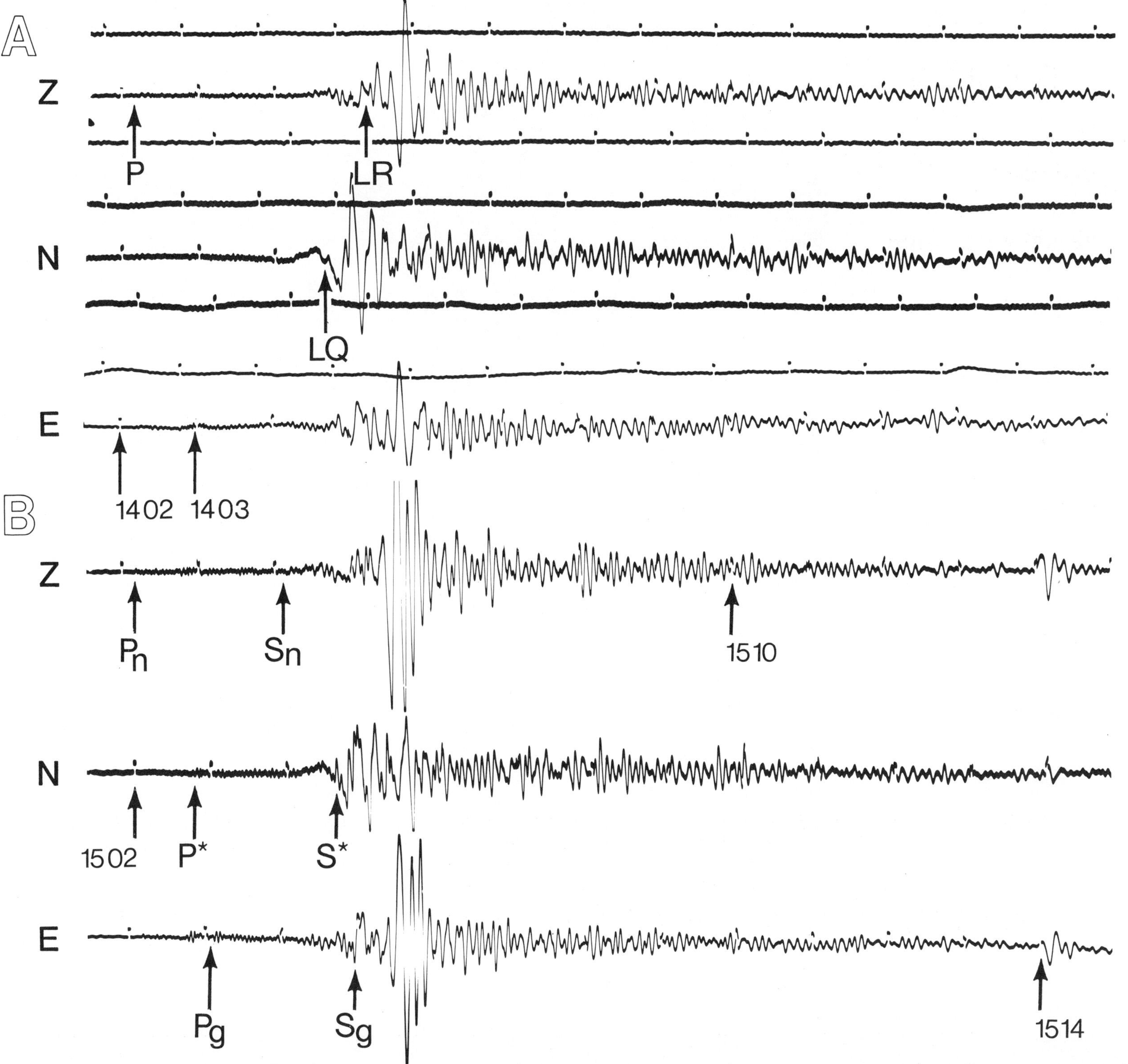
A
Z
P
LR
N
LQ
E
14 02
14 03
B
Z
P_n
S_n
15 10
N
15 02
P*
S*
E
P_g
S_g
15 14

Figure 22

DISTANCE:	11°
RECORDING STATION:	GOL
INSTRUMENTS:	LP, Z, N, E, and Wood-Anderson strong-motion N, E
LOCATION:	California-Mexico Border (Imperial Valley)
COORDINATES:	32.64 N, 115.33 W
DATE:	15 October 1979
TIME:	23 16 55.1 UTC
DEPTH:	7 km
MAGNITUDE:	6.8 M_s (USGS)

COMMENTS:

Compare the Wood-Anderson strong-motion records in this illustration with those of the Berkeley strong-motion torsion seismometer recording (Figure 3).

All of the earthquake was recorded, despite differing characteristics of the long-period and the Wood-Anderson records. Only the early phases were recorded on the long-period components. The surface waves were off scale.

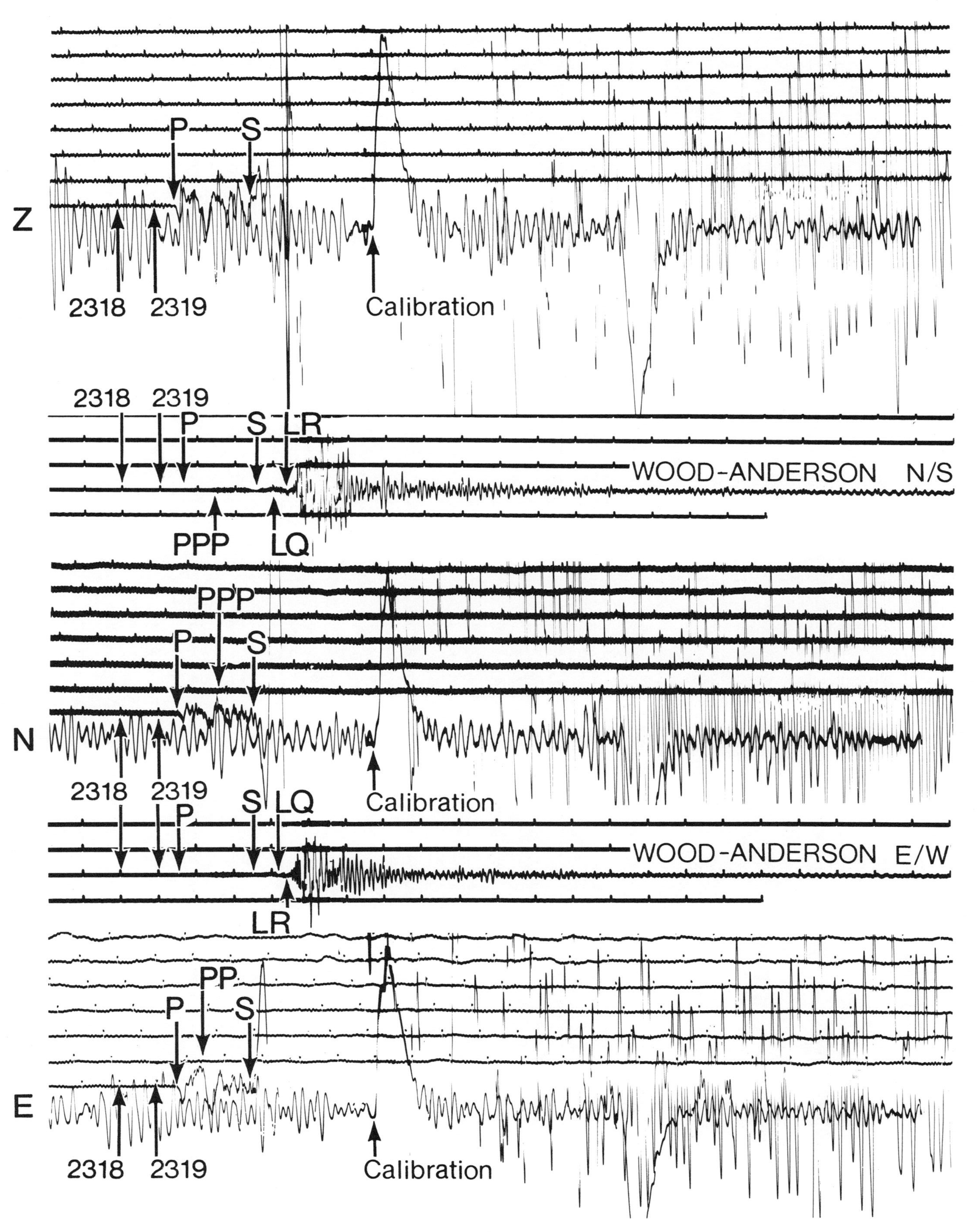

Z
P S
2318 2319
Calibration
2318 2319
P S LR
WOOD-ANDERSON N/S
PPP LQ
PPP
P S
N
2318 2319
P S LQ
Calibration
WOOD-ANDERSON E/W
LR
PP
P S
E
2318 2319
Calibration

Figure 23

DISTANCE: 13°

RECORDING STATION: GOL

INSTRUMENTS: SP Z, N, E

LOCATION: Gulf of California, Baja

COORDINATES: 27.90 N, 111.30 W

DATE: 21 May 1967

TIME: 07 18 13.0 UTC

DEPTH: 33 km

MAGNITUDE: 4.7 M_b (USGS), 5.5 M_s (GOL), 4.4 to 4.8 M_b (BRK)

COMMENTS:

This earthquake is at the distance range between regional and teleseismic distance, where two interpretations might be made, depending upon the recording examined. The long-period components shown on Figure 24 should be noted and compared with these short-period components.

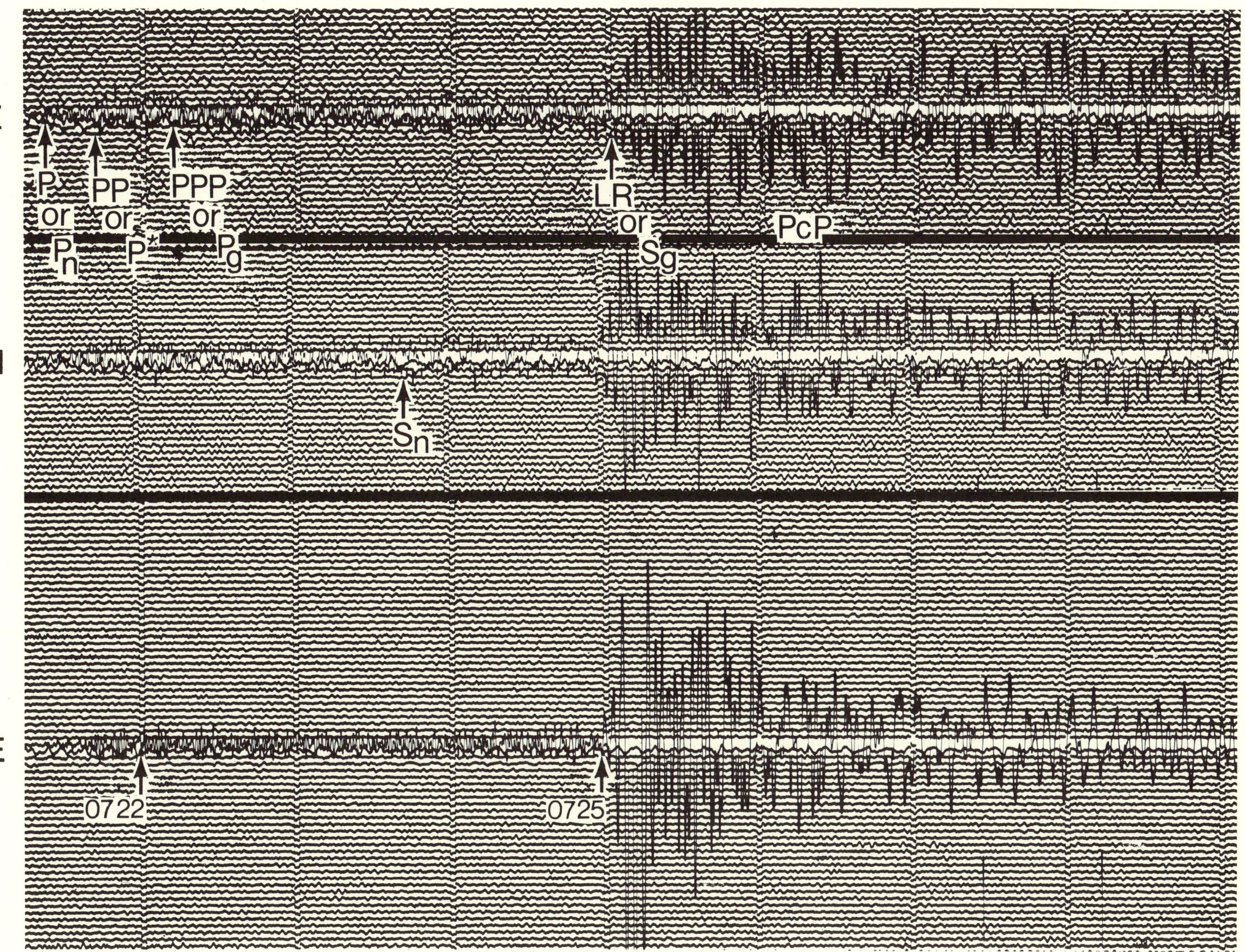

Z
N
E
P
or
Pn
PP
or
P*
PPP
or
Pg
Sn
LR
or
Sg
PcP
0722
0725

Figure 24

DISTANCE: 13°

RECORDING STATION: GOL

INSTRUMENTS: LP Z, N, E

LOCATION: Gulf of California, Baja

COORDINATES: 27.90 N, 111.30 W

DATE: 21 May 1967

TIME: 07 18 13.0 UTC

DEPTH: 33 km

MAGNITUDE: 5.5 M_S (GOL)

COMMENTS:

These records have been reduced 40 percent; 1 mm = 10 seconds. The remarkable Love wave dispersion should be noted, as well as the unusual pulse-like PcP on the long-period vertical, which is marked on Figure 23, but is not so obvious there.

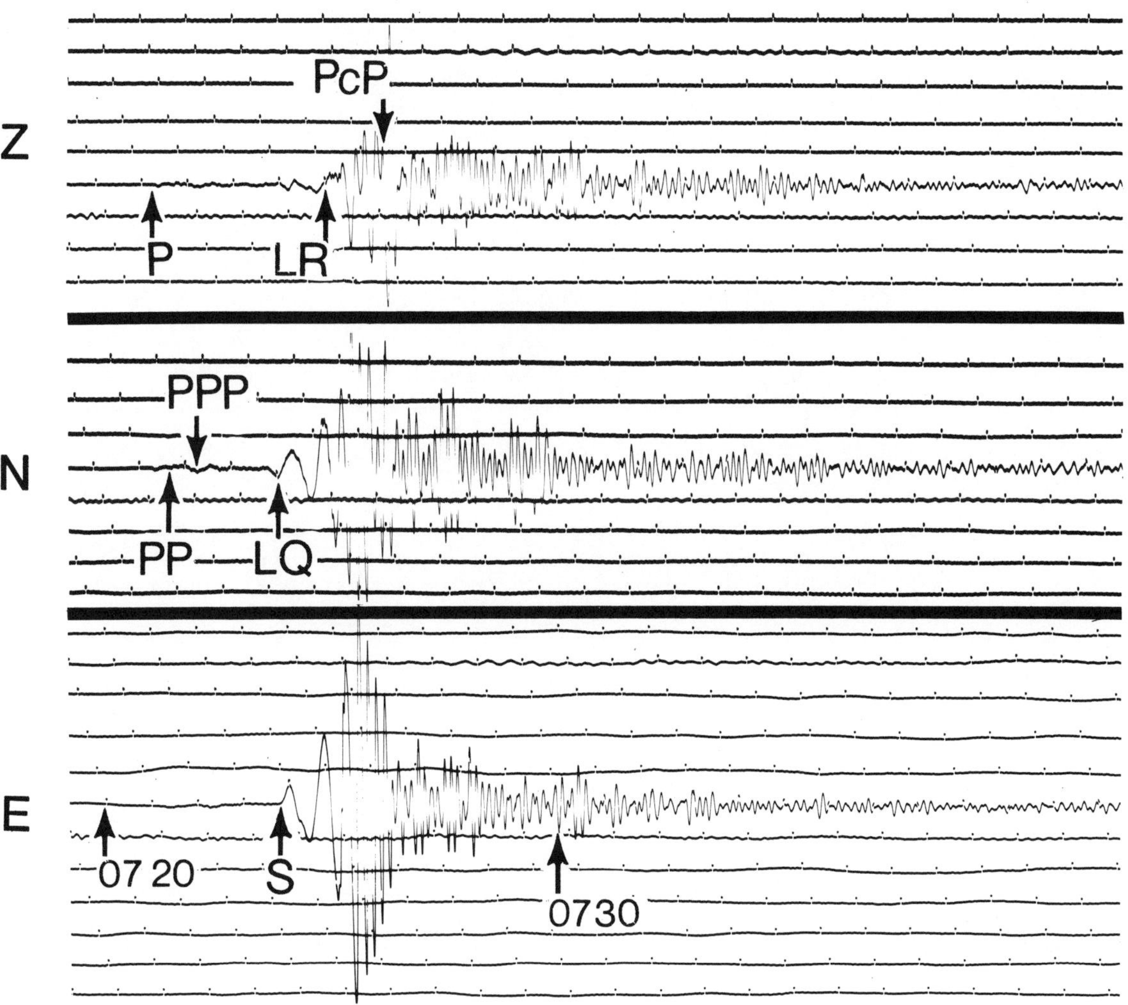

PcP
Z
P
LR
PPP
N
PP
LQ
E
0720
S
0730

Figure 25

DISTANCE: 16°

RECORDING STATION: GOL

INSTRUMENTS: SP Z, N, E

LOCATION: Offshore Northern California (Eureka)

COORDINATES: 40.50 N, 124.60 W

DATE: 10 December 1967

TIME: 12 06 50.3 UTC

DEPTH: 5 km

MAGNITUDE: 5.8 M_S (USGS), 5 1/4 to 5 1/2 M_S (PAS, GOL), 5.5 to 5.7 M_S (BRK), 5 3/4 to 6 M_S (PAL)

COMMENTS:

This earthquake was felt strongly at Eureka, Arcata, and Rio Dell.

The very large P wave and relatively small S may be characteristic of an oceanic path, but this earthquake also produced higher-mode continental Rayleigh waves, which can be seen on both the short-period records and long-period records in Figure 26.

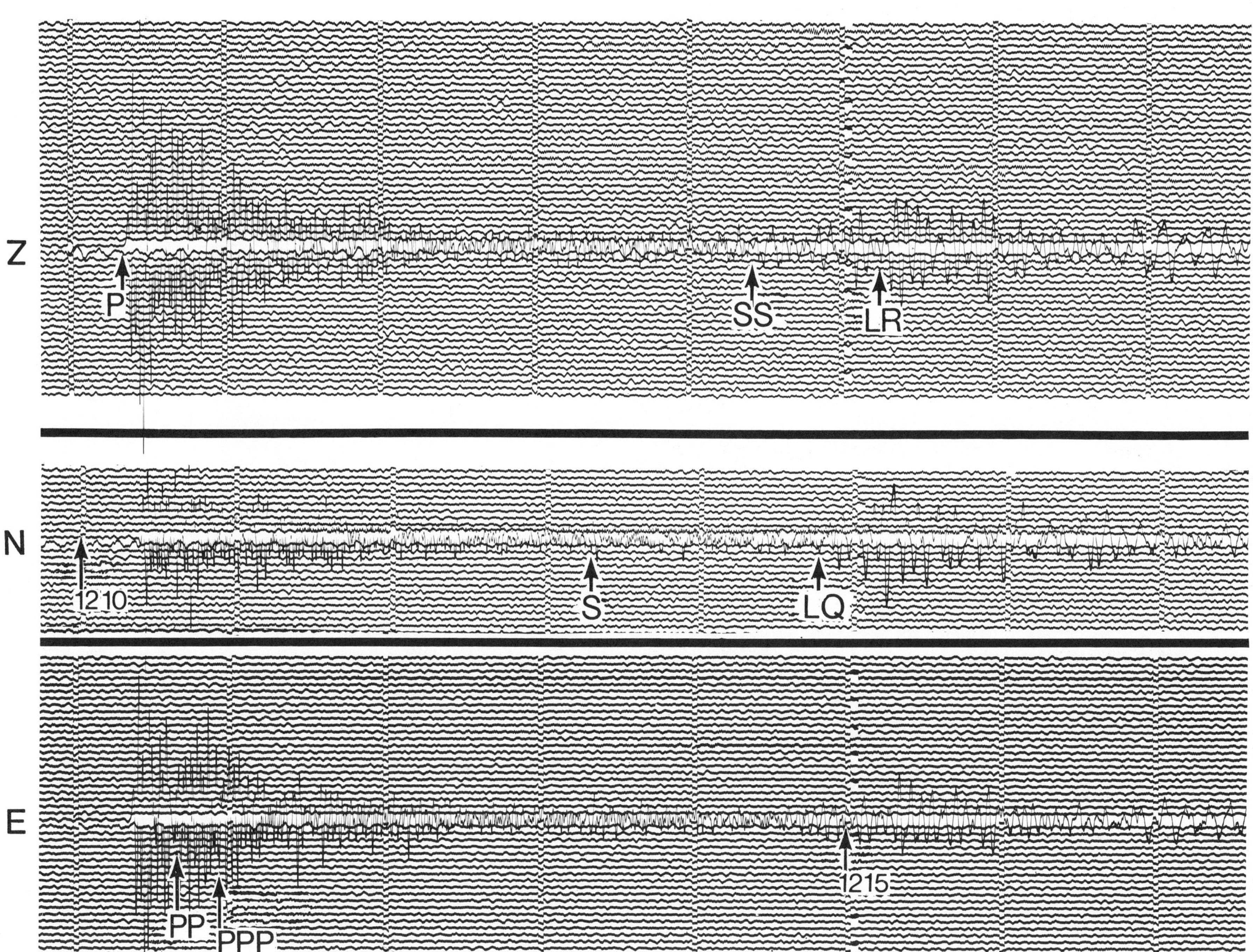

Z
P
SS
LR
N
1210
S
LQ
E
PP
PPP
1215

Figure 26

DISTANCE:	16°
RECORDING STATION:	GOL
INSTRUMENTS:	LP Z, N, E
LOCATION:	Offshore Northern California (Eureka)
COORDINATES:	40.50 N, 124.60 W
DATE:	10 December 1967
TIME:	12 06 50.3 UTC
DEPTH:	5 km
MAGNITUDE:	5 1/4 to 5 1/2 M_S (GOL)

COMMENTS:

This is the same earthquake as shown in Figure 25. There are unusually large and well dispersed surface waves of several modes for such a short distance.

These records are reduced 40 percent; 1 mm = 10 seconds.

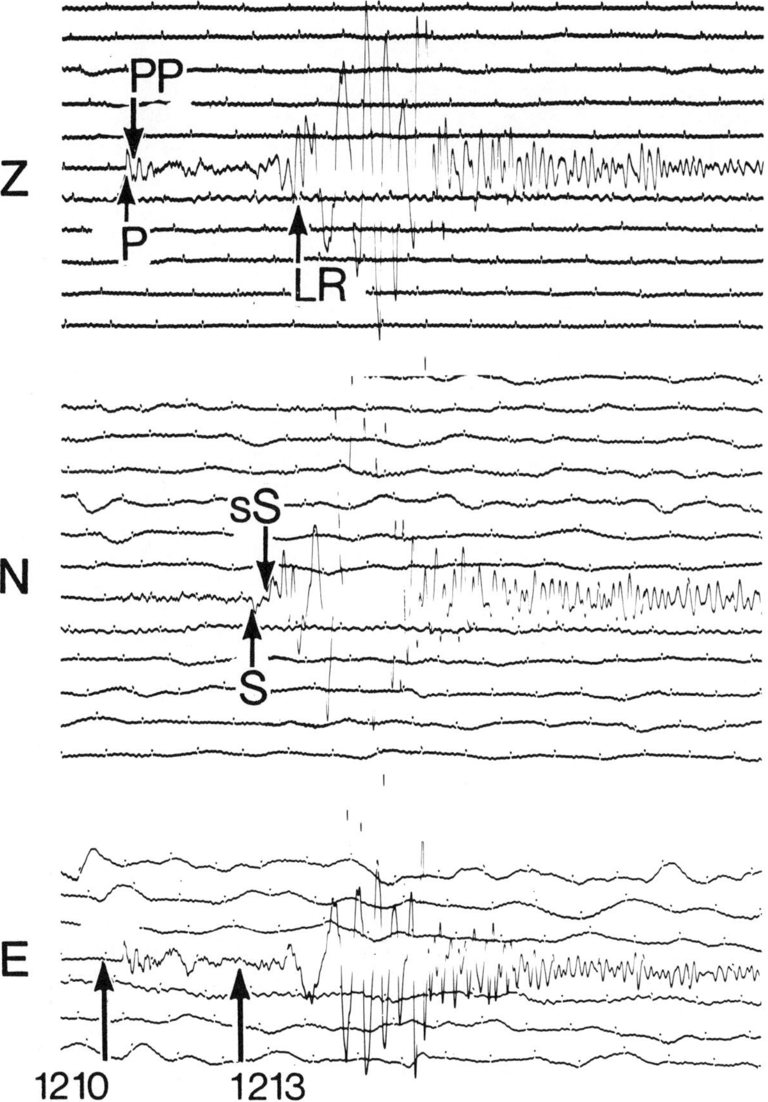

PP
Z
P
LR
sS
N
S
E
1210
1213

Figure 27

	A	B
DISTANCE:	19°	23°
RECORDING STATION:	GOL	GOL
INSTRUMENTS:	LP Z, N, E	LP Z, N, E
LOCATION:	Revilla Gigedo Islands	Vera Cruz, Mexico
COORDINATES:	21.60 N, 108.80 W	19.10 N, 95.80 W
DATE:	2 March 1967	11 March 1967
TIME:	13 21 44.7 UTC	14 44 59.2 UTC
DEPTH:	33 km	33 km
MAGNITUDE:	4.6 M_S (GOL)	5.5 M_S (GOL)

COMMENTS:

The second earthquake (B) injured 3 persons and caused moderate property damage at Vera Cruz and Baca del Rio. It was also felt at Jalapa.

These two events were within 4° of each other, at almost the same azimuth and the same depth, but the second was one magnitude larger. In both earthquakes, the Love waves are most prominent on the E-W components. The differences in the long-period waves are more noteworthy than are their similarities.

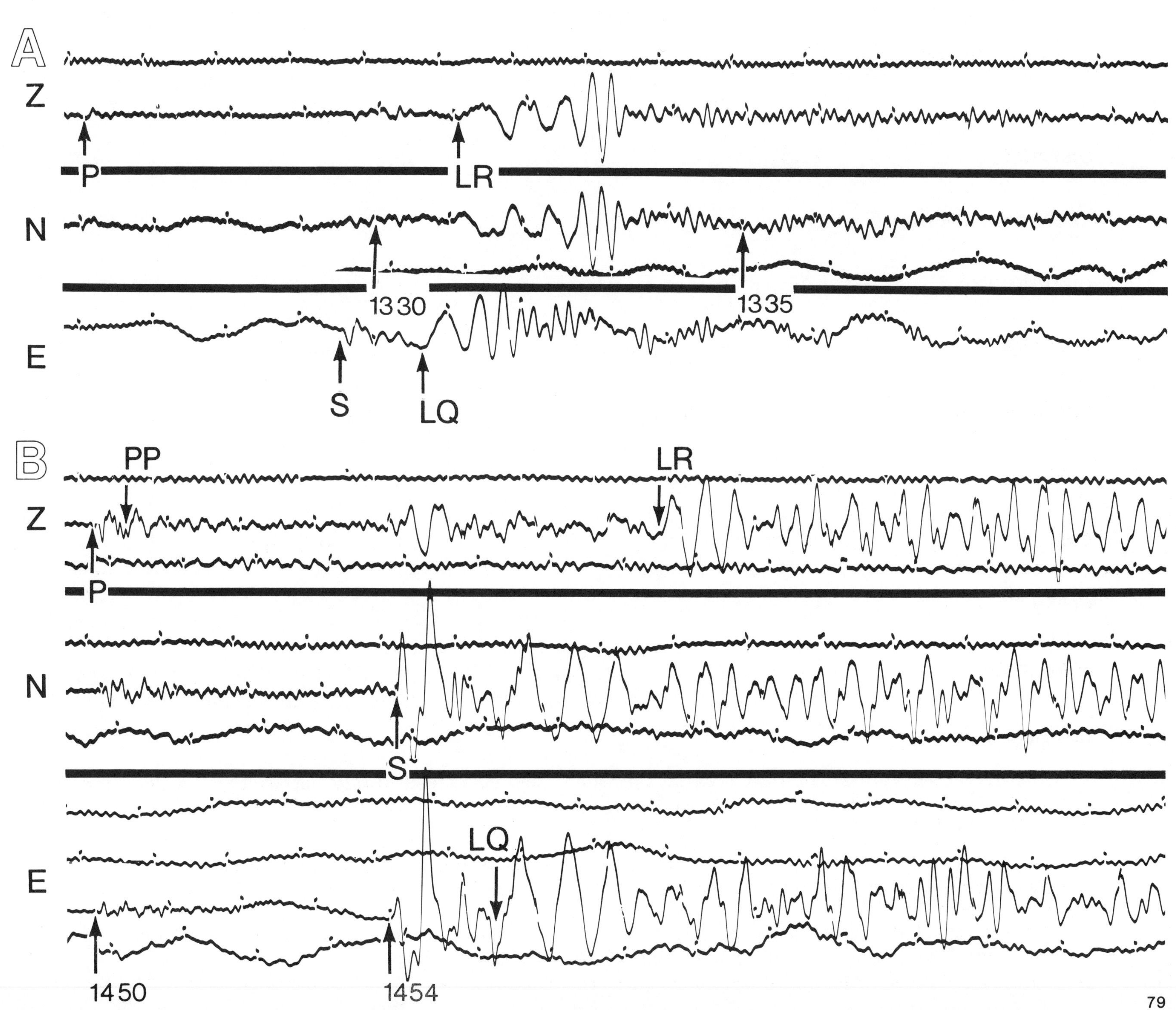

A
Z
P
LR
N
13 30
1335
E
S
LQ
B
PP
LR
Z
P
N
S
LQ
E
14 50
1454

Figure 28

	A	B
DISTANCE:	29.4°	30.2°
RECORDING STATION:	PAL	PAL
INSTRUMENTS:	LP Z, N, E	LP Z, N, E
LOCATION:	North Atlantic Ocean	El Salvador
COORDINATES:	52.90 N, 34.20 W	13.50 N, 89.30 W
DATE:	5 July 1965	3 May 1965
TIME:	08 31 58.9 UTC	10 01 35.2 UTC
DEPTH:	33 km	23 km
MAGNITUDE:	5.7 M_S (PAL), 5 3/4 M_S (PAS), 5.8 M_S (BRK), 5.6 M_S (USGS)	6 to 6 1/4 M_S (PAL), 6 1/4 M_S (PAS), 5 1/2 to 6 1/2 M_S (BRK), 5.1 M_S (USGS)

COMMENTS:

The dispersed wave train in A from S, through SS and LQ is especially noteworthy. B has similar wave patterns. These types of waves travel an oceanic path from the NE (A), as well as a continental path from the SSW, as in B.

The El Salvador earthquake killed 125 persons, injured 400, and wrecked 4000 homes in San Salvador Cisneros District, San Maros, and Santo Thomas.

Love waves of different periods appear on the horizontal components during the same time interval that Rayleigh waves may be seen on the vertical component.

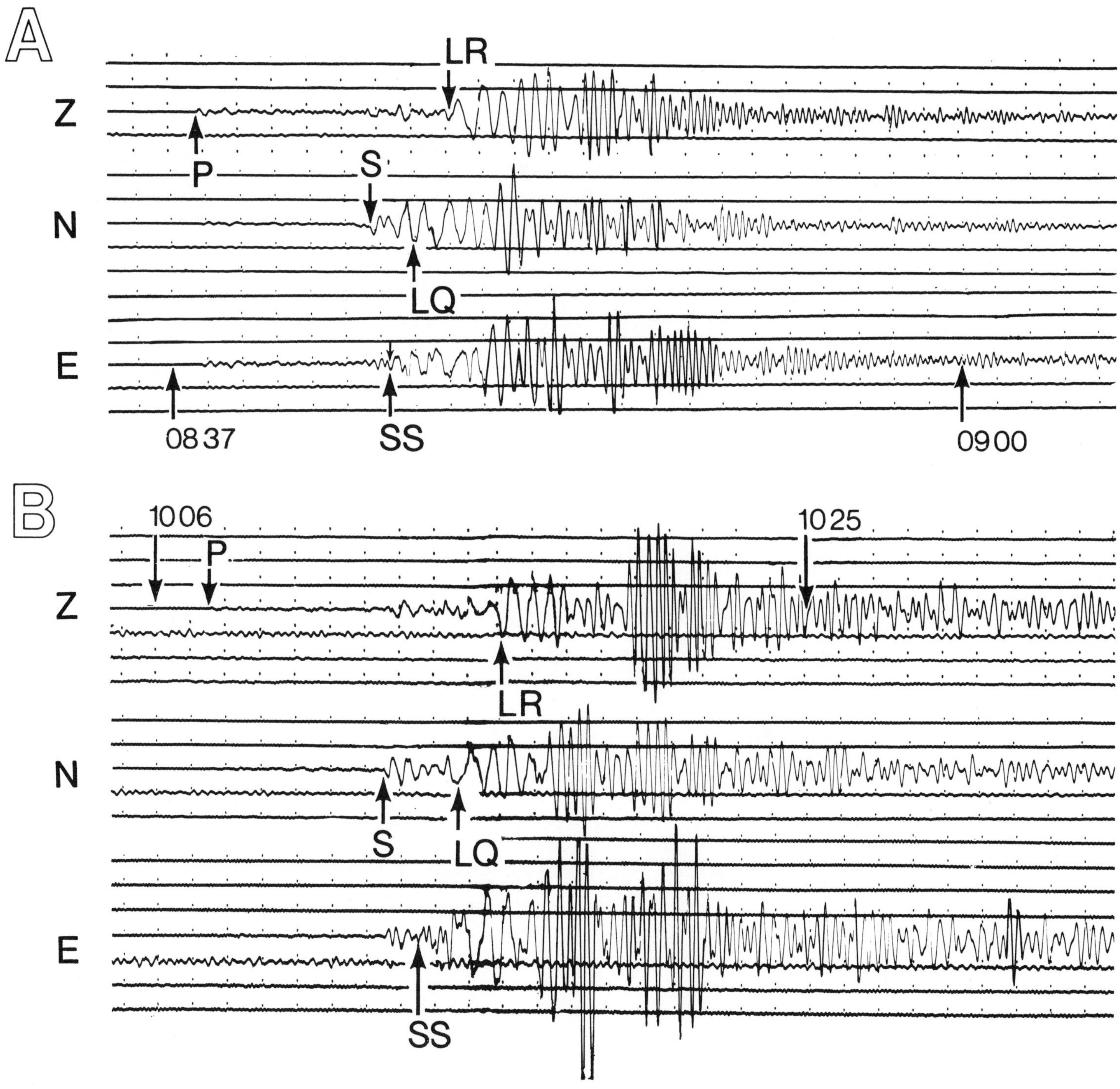

A
Z
N
E
LR
P
S
LQ
SS
08 37
09 00
B
Z
N
E
10 06
10 25
P
LR
S
LQ
SS

Figure 29

	A	B
DISTANCE:	30°	35.5°
RECORDING STATION:	PAL	PAL
INSTRUMENTS:	LP Z, N, E	LP Z, N, E
LOCATION:	Offshore Nicaragua	Offshore Central America
COORDINATES:	12.50 N, 87.40 W	6.50 N, 84.40 W
DATE:	20 October 1965	9 September 1965
TIME:	23 54 29.9 UTC	10 02 25.4 UTC
DEPTH:	70 km	27 km
MAGNITUDE:	5.4 M_S (USGS), 5 to 5.5 M_S (BRK), 5 3/4 M_S (PAL),	5.5 M_S (USGS), 6.75 to 7 M_S (BRK), 6 3/4 M_S (PAS), 6 1/2 to 6 3/4 M_S (PAL)

COMMENTS:

Core-grazing reflected phases may be interpreted either as guided layered surface waves or higher modes of continental Rayleigh waves.

Earthquake A was felt in western Nicaragua and El Salvador.

Earthquake B is unusual in that the S coupled P_L phase and the multiple SS appear so prominently on the vertical component.

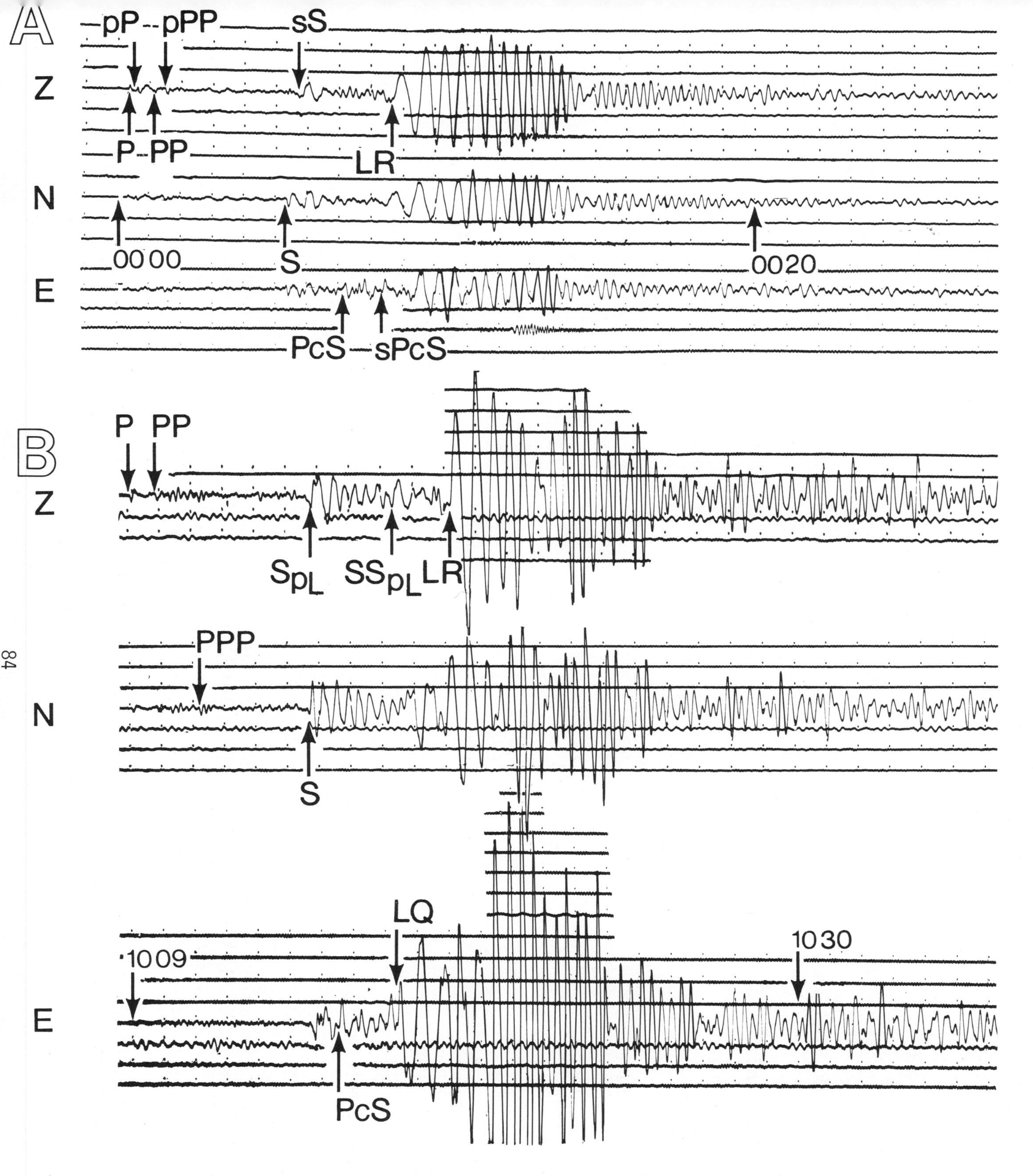

A
Z
pP--pPP
sS
P--PP
LR
N
00 00
S
0020
E
PcS--sPcS
B
P PP
Z
SpL SSpL LR
PPP
N
S
LQ
1009
1030
E
PcS

Figure 30

	A	**B**
DISTANCE:	34°	34°
RECORDING STATION:	GOL	GOL
INSTRUMENTS:	LP Z, N, E	SP Z, N, E
LOCATION:	Gulf of Alaska	Gulf of Alaska
COORDINATES:	56.20 N, 149.90 W	56.20 N, 149.90 W
DATE:	2 August 1964	2 August 1964
TIME:	08 36 16.9 UTC	08 36 16.9 UTC
DEPTH:	31 km	31 km
MAGNITUDE:	5.4 M_b (USGS)	5.4 M_b (USGS)

COMMENTS:

The early phases are more clearly shown on the short-period records (B). The surface waves are the long period, first mode Love and Rayleigh waves. LQ is seen first on the horizontal components; LR comes later, on the vertical component (A). Surface waves may also be seen on the short-period records on the line beneath the P waves, 15 minutes later.

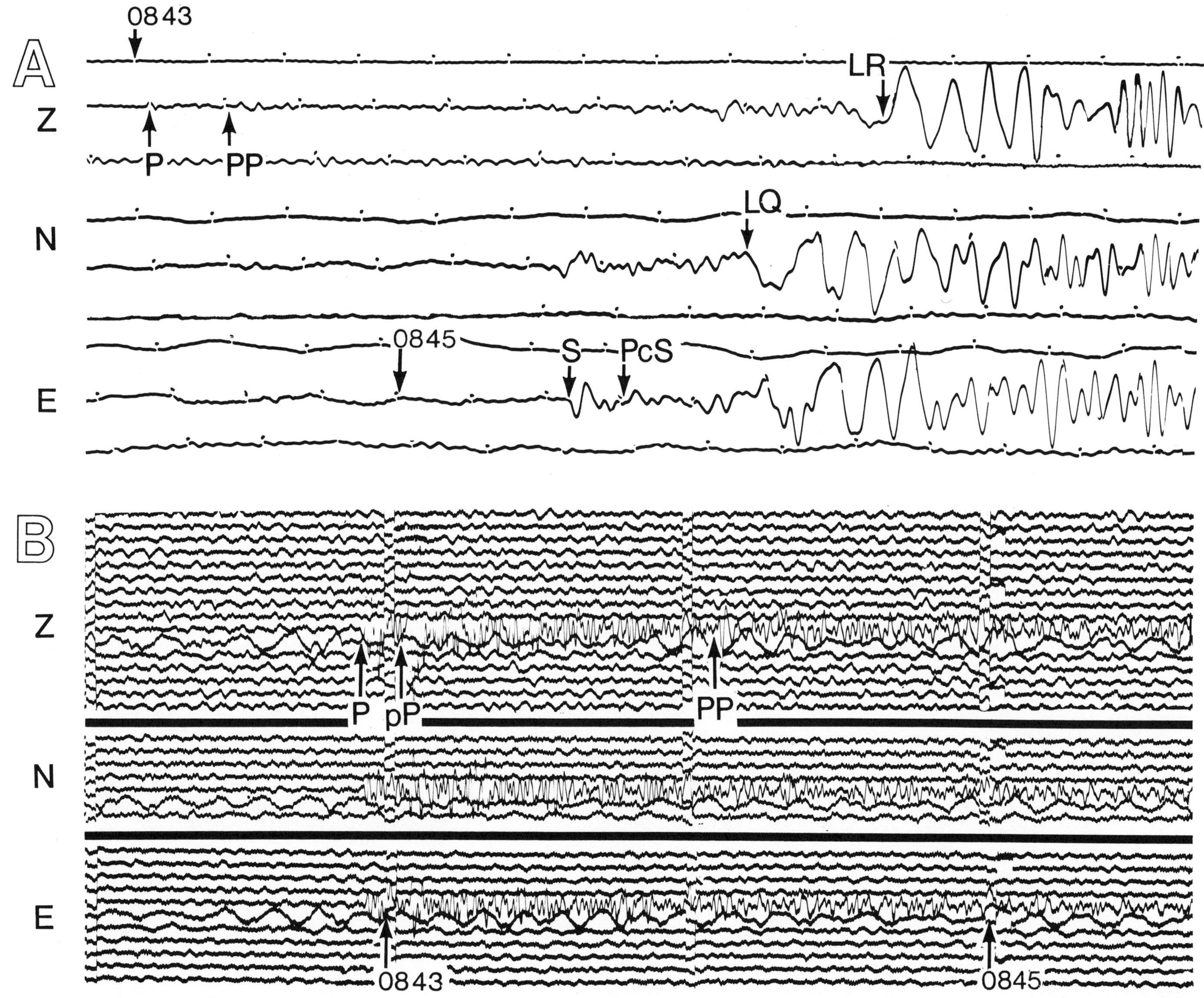

A
Z
0843
P
PP
LR
N
LQ
E
0845
S
PcS
B
Z
P pP
PP
N
E
0843
0845

Figure 31

	A	B
DISTANCE:	39.2°	44.5°
RECORDING STATION:	PAL	PAL
INSTRUMENTS:	LP Z, N, E	LP Z, N, E
LOCATION:	Offshore Northern California	Southeastern Alaska
COORDINATES:	41.50 N, 127.10 W	60.30 N, 141.20 W
DATE:	18 April 1965	27 June 1965
TIME:	06 33 58.8 UTC	11 08 55.0 UTC
DEPTH:	20 km	12 km
MAGNITUDE:	5.6 M_b (USGS), 6.0 M_s (BRK), 5 3/4 to 6 M_s (PAL)	5.3 M_b (USGS), 5.0 M_s (BRK), 5 3/4 to 6 M_s (PAL)
COMMENTS:	Compare A with the records of B in Figure 29, for different S coupled P_L phase appearing on the vertical component. Long-period first mode Love and Rayleigh waves can be seen on A but not on B.	These records illustrate higher-mode surface waves of both Love and Rayleigh types. Lg appears as a Love wave on the bottom trace, the E-W component, and is not seen on the vertical. Lg and Rg often arrive with the pulse-like character shown and are typical of earthquakes having continental paths to the station. Long-period surface waves beneath Lg are unrelated to this earthquake, coming from another, more distant one.

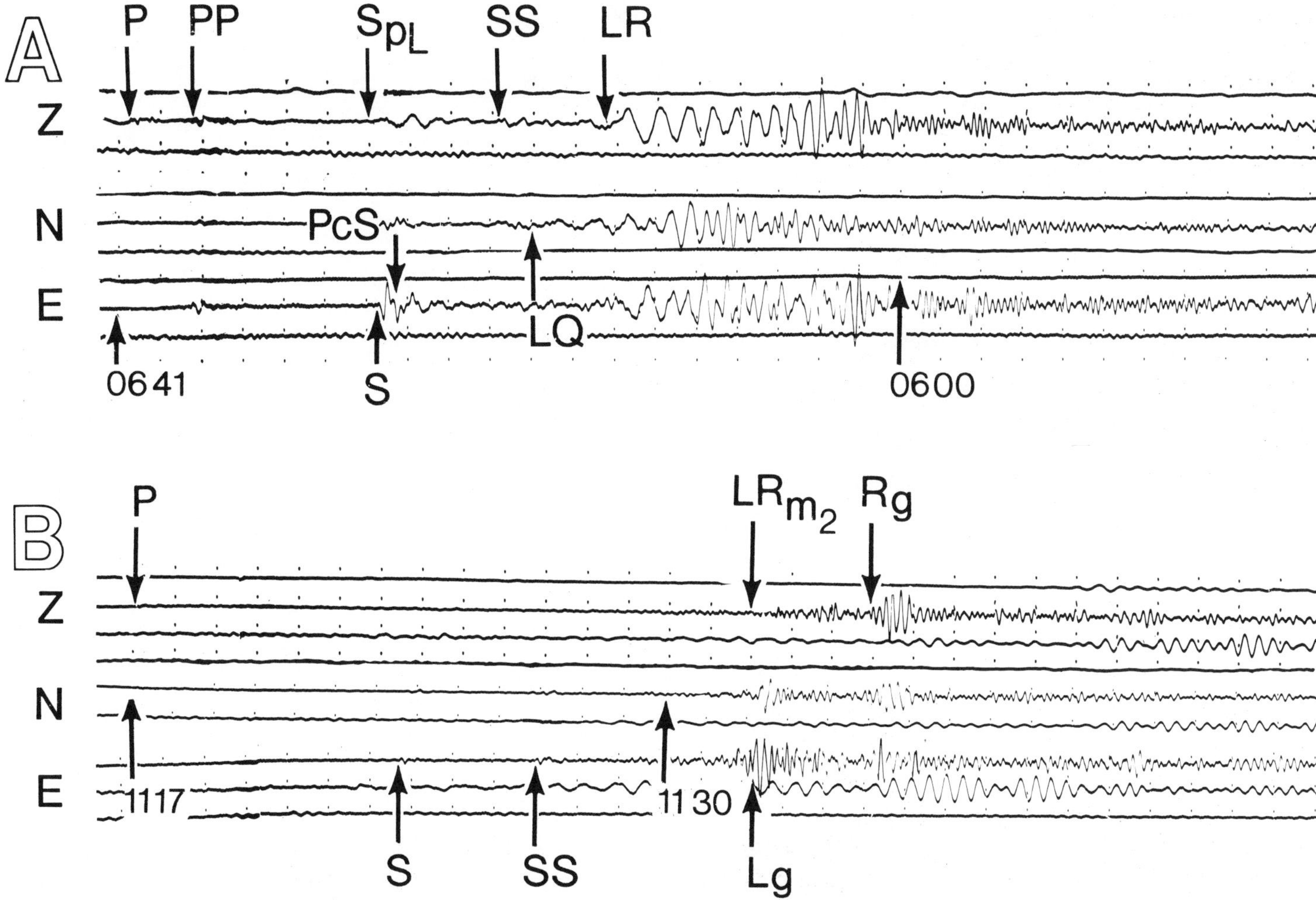

A
P
PP
S_pL
SS
LR
Z
N
PcS
E
LQ
06 41
S
06 00
B
P
LR_m2
Rg
Z
N
E
1117
11 30
S
SS
Lg

Figure 32

DISTANCE: 43°

RECORDING STATION: GOL

INSTRUMENTS: LP Z, N, E

LOCATION: Northern Colombia, South America

COORDINATES: 6.80 N, 73.00 W

DATE: 29 July 1967

TIME: 10 24 24.6 UTC

DEPTH: 161 km

MAGNITUDE: 6.0 M_b (USGS), 6 1/2 M_s (PAL), 6 1/2 to 6 3/4 M_s (PAS), 6.1 to 6.3 M_s (BRK)

COMMENTS:

This earthquake killed 10 persons and caused moderate property damage. It was felt all over Colombia and western Venezuela. P waves were recorded by 144 stations around the world.

The first motion of P is a clear dilatation from the southeast. The SS and LR are pulse-like, possibly the effect of the depth of focus. Surface waves attenuate rapidly at intermediate focal depths.

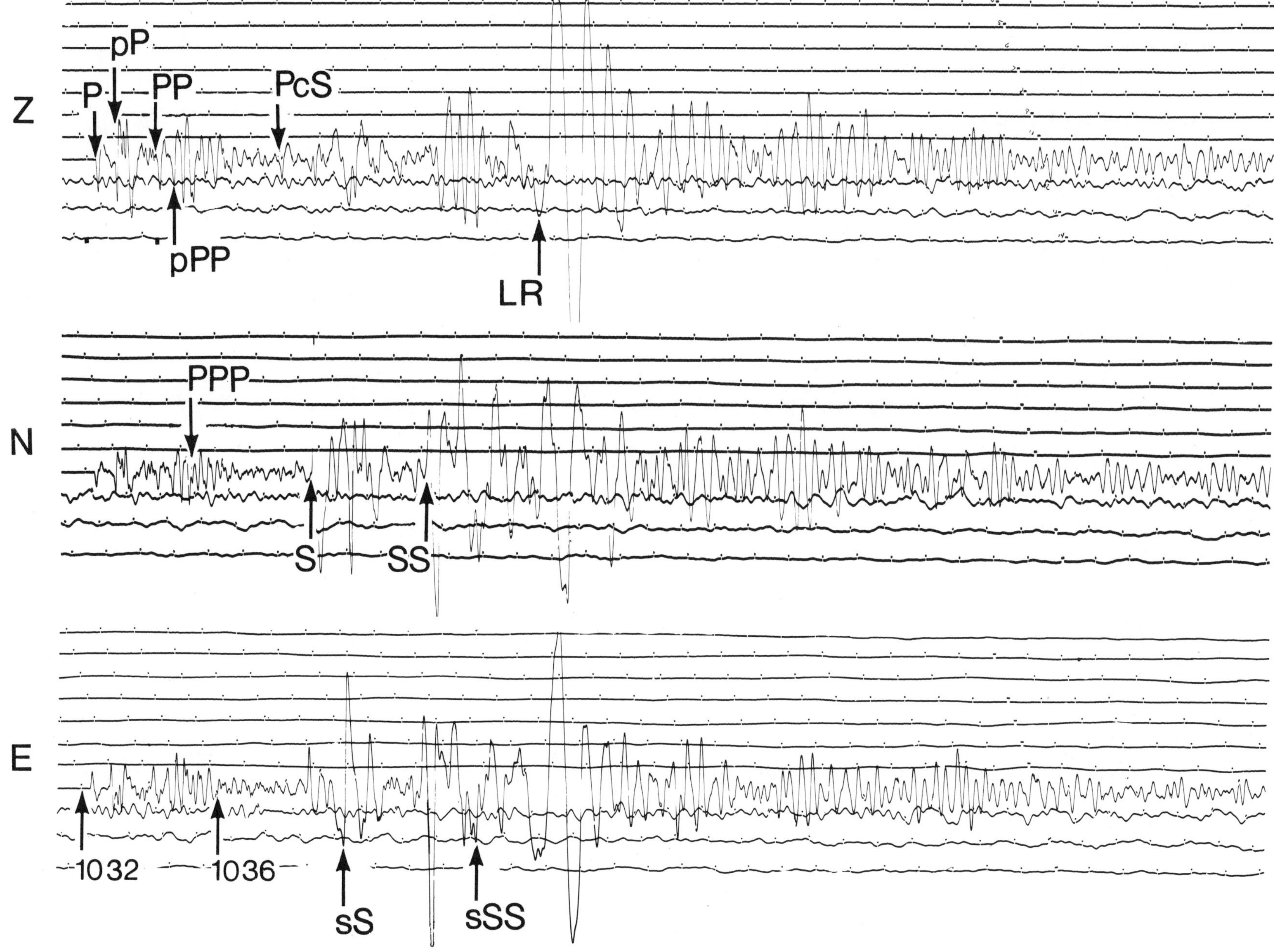

89

Figure 33

DISTANCE:	51.3°
RECORDING STATION:	PAL
INSTRUMENTS:	LP Z, N, E
LOCATION:	Kodiak Island, Alaska
COORDINATES:	56.60 N, 152.90 W
DATE:	23 June 1965
TIME:	11 09 15.3 UTC
DEPTH:	36 km
MAGNITUDE:	6 1/2 to 6 3/4 M_S (PAL), 6 1/4 to 6 1/2 M_S (PAS), 5 3/4 M_S (BRK)

COMMENTS:

The constant frequency waves from P to S are unusual. S coupled P_L waves are unusual. The SS coupled P_L, with possibly more multiples of S on the Z component up to LR, are also unusual. The path is continental, but this is not the typical wave pattern for a continental shock. Compare these records with those of the Svalbard earthquake in Figure 34.

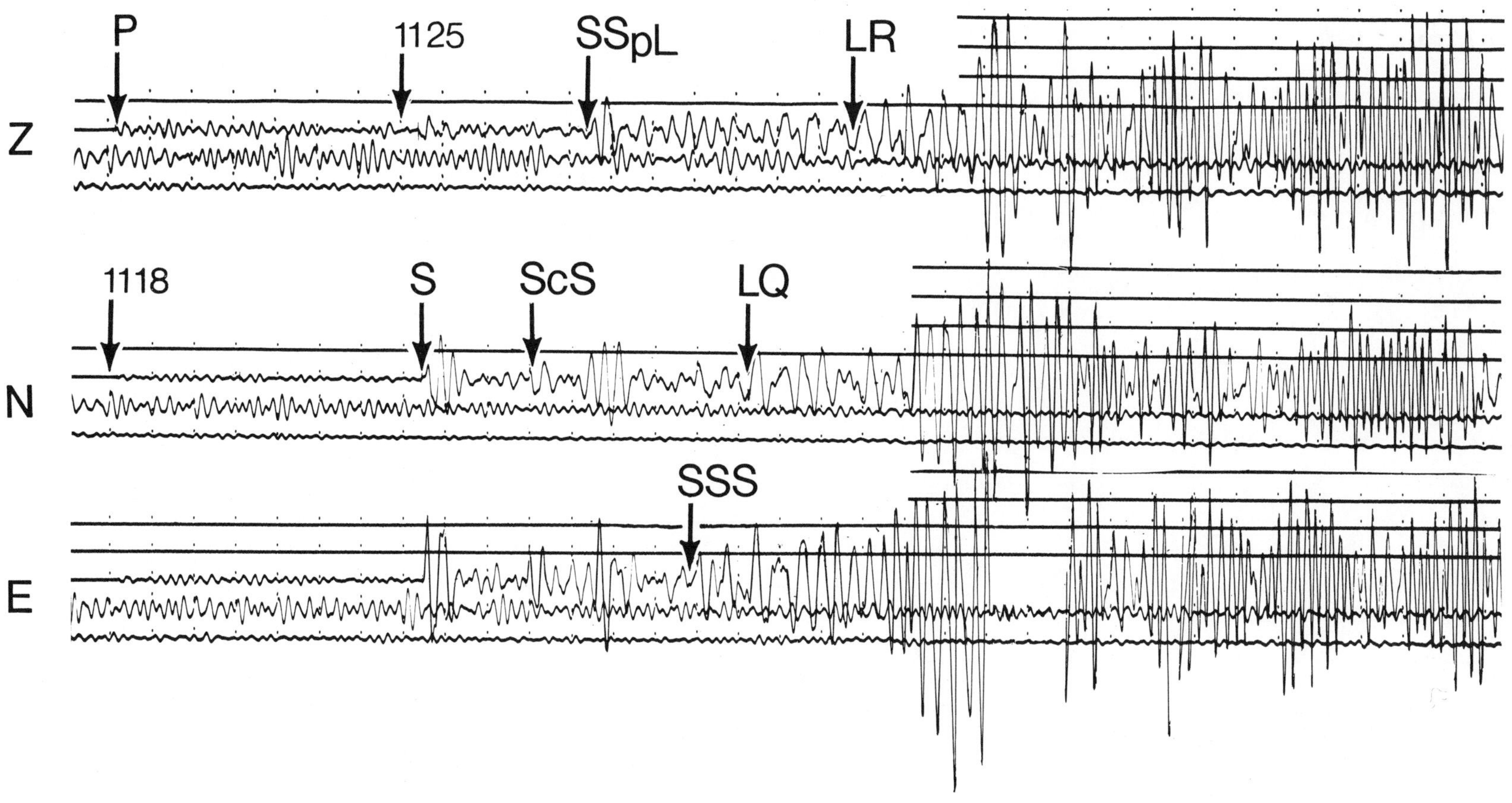

P
1125
SSpL
LR
Z
1118
S
ScS
LQ
N
SSS
E

Figure 34

DISTANCE:	54°
RECORDING STATION:	GOL
INSTRUMENTS:	LP Z, N, E
LOCATION:	North of Svalbard
COORDINATES:	80.20 N, 1.00 W
DATE:	23 November 1967
TIME:	13 42 01.6 UTC
DEPTH:	10 km
MAGNITUDE:	5.8 M_b (USGS), 6 1/4 M_s (PAS), 6 3/4 to 7 M_s (GOL)

COMMENTS:

Higher modes of surface waves (periods of 20 to 12 seconds), beginning as early as SS, are seen on all components. The same frequencies of lower amplitude are also seen after all the body phases. Seismic waves from this earthquake traversed a largely continental path through Greenland and the Canadian shield.

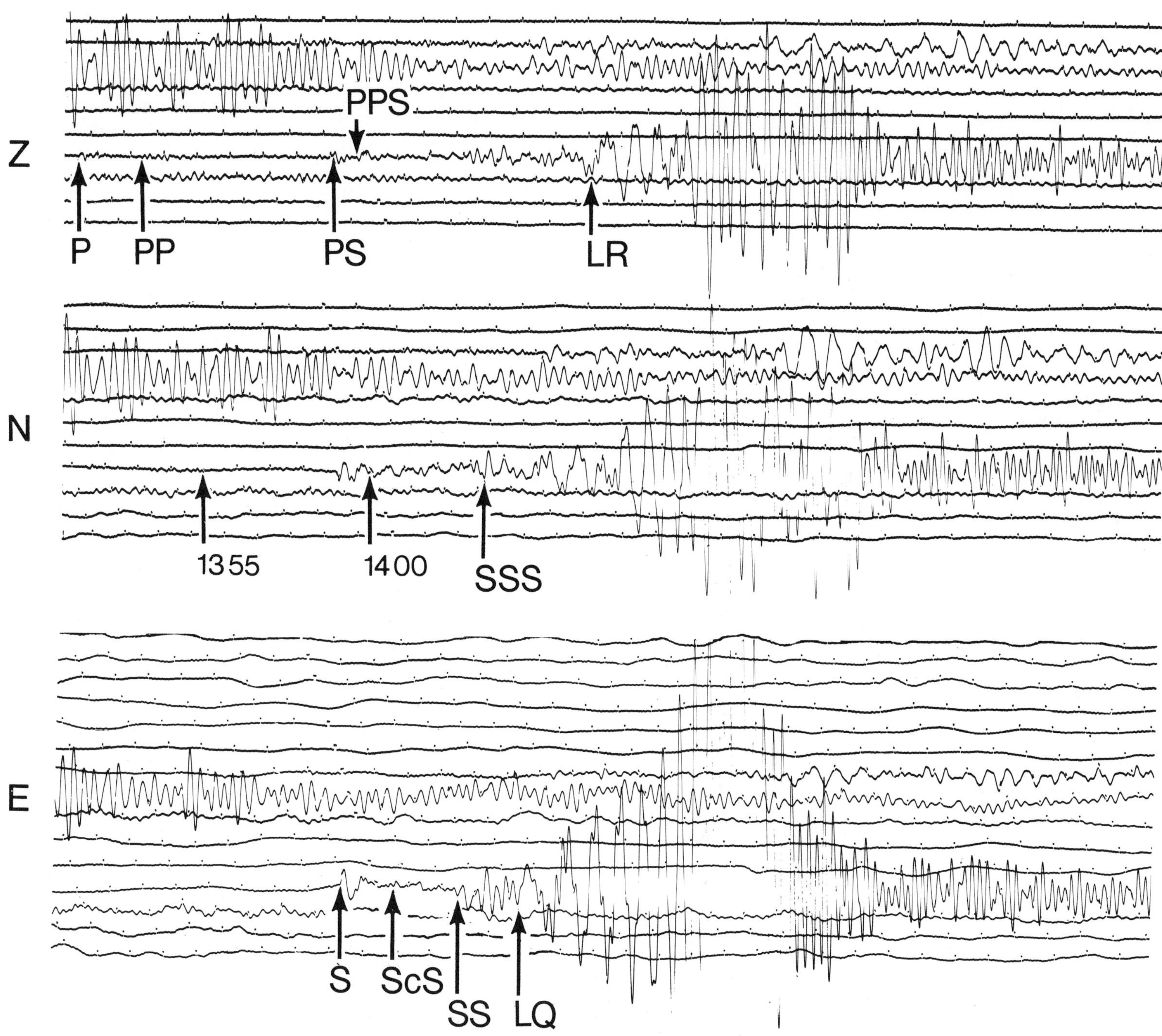
Z
PPS
P PP PS LR
N
13 55 14 00 SSS
E
S ScS SS LQ

Figure 35

	A	**B**
DISTANCE:	59.8°	70.8°
RECORDING STATION:	PAL	PAL
INSTRUMENTS:	LP Z, N, E	LP Z, N, E
LOCATION:	Northern Chile	Greece
COORDINATES:	19.00 S, 69.10 W	38.70 N, 22.60 E
DATE:	20 August 1965	6 July 1965
TIME:	09 42 48.5 UTC	03 18 44.6 UTC
DEPTH:	129 km	28 km
MAGNITUDE:	6.0 M_b (USGS), 5.7 M_b (BRK), 6 1/2 M_s (PAS)	5.9 M_b (USGS), 6 M_s (BRK) 6 1/4 M_s (PAS), 6 to 6 1/4 M_s (PAL)

COMMENTS:

A

This earthquake was felt at Arequipa, Peru, and at Arica and and Iquique, Chile.

The small P and large core reflections after S are often seen at intermediate focal depths.

B

One person was reported killed and six persons were injured at Eratini. There was considerable damage in northern Peloponnesus. A seismic sea wave was reported at Eratini.

Compare the size of P with that of A for events of approximately the same magnitude. The larger P for the shallower event is not usual. Compare the wave forms for LQ and LR on these records with those of Figure 34. The differences are largely due to the difference in azimuths, as well as the paths to the station.

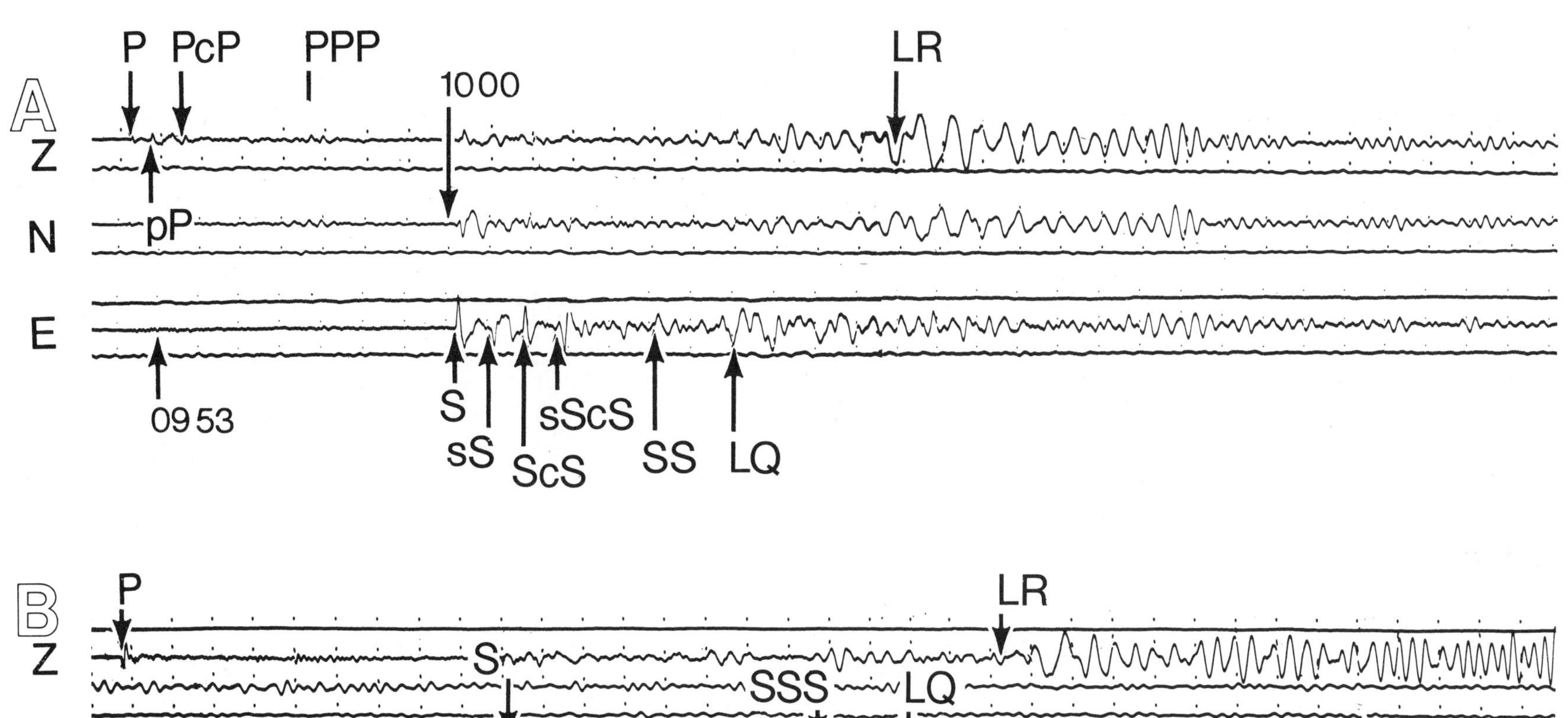

P
PcP
PPP
I
LR
1000
A
Z
N
pP
E
0953
S
sS
ScS
sScS
SS
LQ
P
LR
B
Z
S
SSS
LQ
N
SS
ScS
E
PcS
0350
0400

Figure 36

	A	**B**
DISTANCE:	78°	78°
RECORDING STATION:	GOL	GOL
INSTRUMENTS:	SP Z, N, E	LP Z, N, E
LOCATION:	Santiago del Estero Province, Argentina	Santiago del Estero Province, Argentina
COORDINATES:	27.40 S, 63,30 W	27.40 S, 63.30 W
DATE:	17 January 1967	17 January 1967
TIME:	01 07 54.3 UTC	01 07 54.3 UTC
DEPTH:	590 km	590 km
MAGNITUDE:	5.5 M_b (USGS), 6 1/4 M_b (PAS), 4.8 to 5.2 M_b (BRK)	5.5 M_b (USGS), 6 1/4 M_b (PAS), 4.8 to 5.2 M_b (BRK)

COMMENTS:

The long separation in time between P and pP indicates deep focus. The second phase on the short-period records is pP, not to be confused with PP. Absence of surface waves is another indication of great depth.

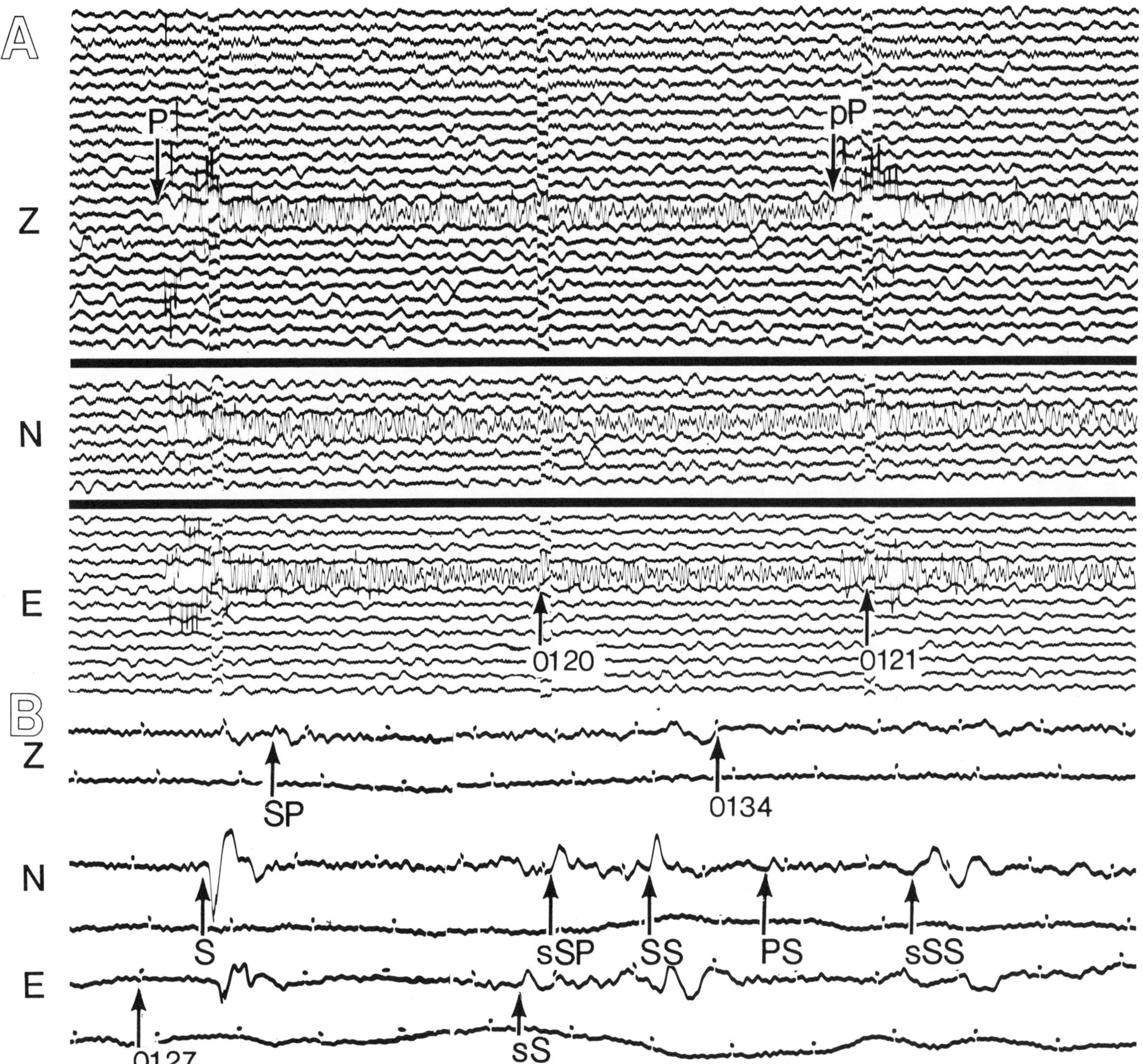

A
Z
P
pP
N
E
0120
0121
B
Z
SP
0134
N
S
sSP
SS
PS
sSS
E
0127
sS

Figure 37

DISTANCE: 81°

RECORDING STATION: GOL

INSTRUMENTS: LP Z, N, E

LOCATION: Honshu, Japan

COORDINATES: 38.30 N, 142.10 E

DATE: 17 January 1967

TIME: 11 59 31.5 UTC

DEPTH: 60 km

MAGNITUDE: 5.9 M_s, 5.9 M_b (USGS), 6 1/2 (PAS), 6.5 (BRK)

COMMENTS:

The body phases are of relatively high amplitude for a shock at this depth. First mode Love and Rayleigh waves of 60-second periods are well developed on these records.

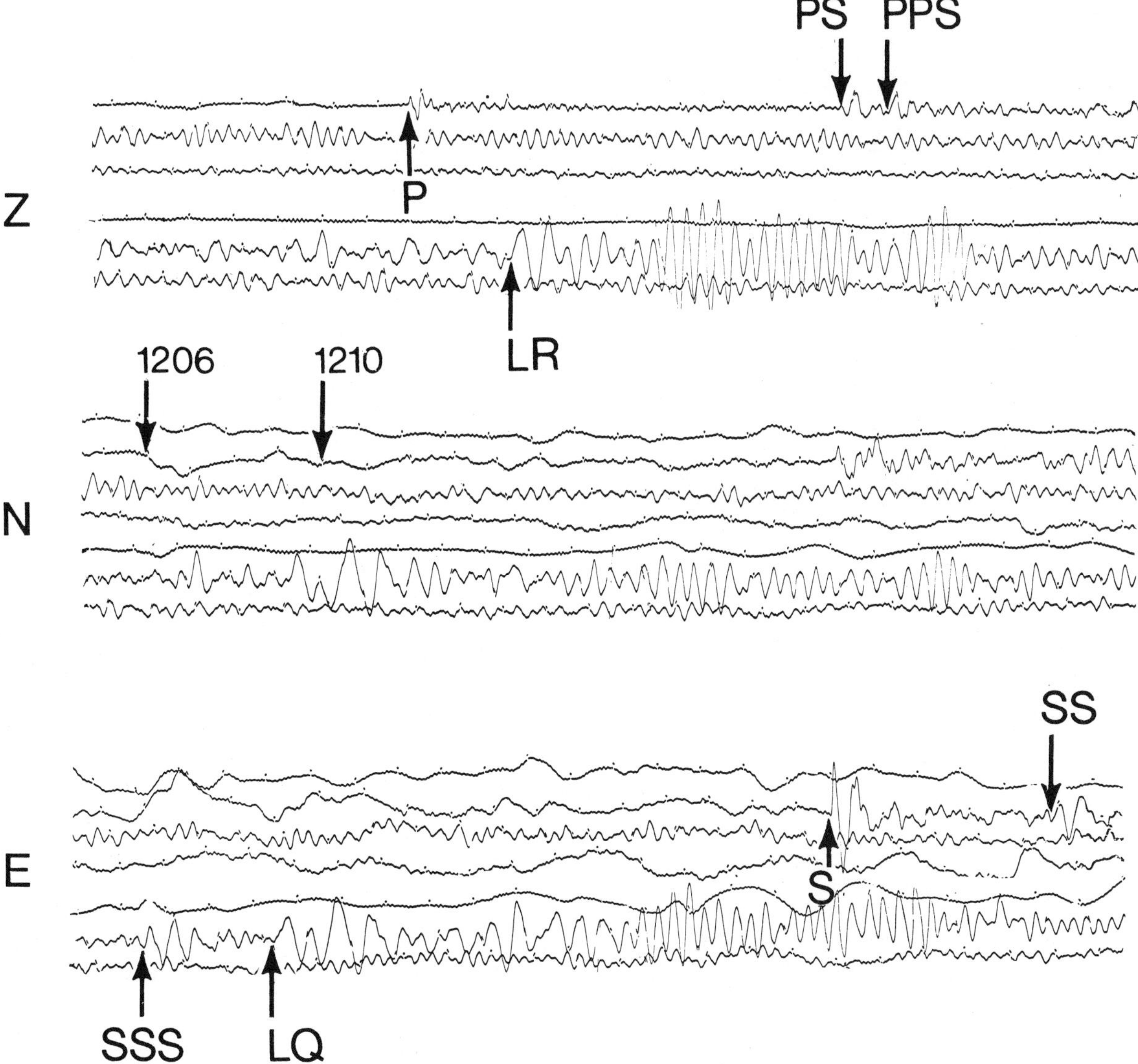

PS PPS
Z
P
LR
1206 1210
N
SS
E
S
SSS LQ

Figure 38

DISTANCE: 85.5°

RECORDING STATION: GOL

INSTRUMENTS: LP Z, N, E

LOCATION: Albania

COORDINATES: 41.50 N, 20.50 E

DATE: 30 November 1967

TIME: 07 23 57.5 UTC

DEPTH: 29 km

MAGNITUDE: 6 1/2 to 6 3/4 M_S (GOL), 6.0 M_b (USGS)

COMMENTS:

This earthquake killed 19 persons and left 50,000 homeless through the Yugoslavia-Albania border area. Hardest hit was Debar, Yugoslavia, where every building was left uninhabitable. Debar is 60 miles southwest of Skopje, the city that was demolished in July, 1963 by an earthquake of magnitude 5 3/4 to 6, which caused 1000 fatalities.

The distance of this earthquake from the station makes this the first series of records where SKS can be distinguished from S.

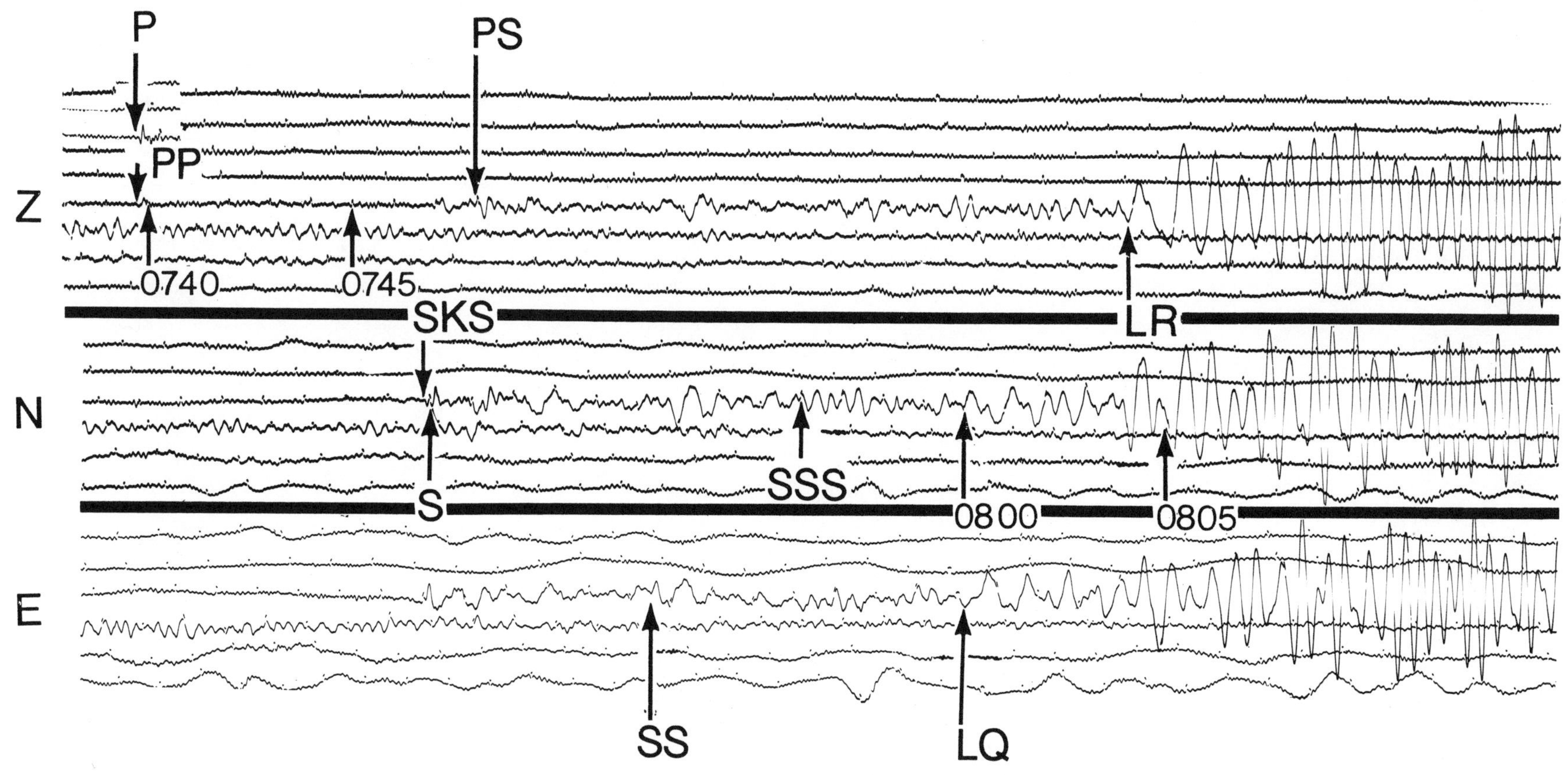

P
PS
PP
Z
0740
0745
SKS
LR
N
S
SSS
0800
0805
E
SS
LQ

Figure 39

	A,B,C	D,E

DISTANCE: 80° to 89°

RECORDING STATION:
A: KPK Kanaka Peak, California
B: TNP Tonapah, Nevada
C: BMN Battle Mt., Nevada
D: ALQ Albuquerque, New Mexico
E: BDW Boulder, Wyoming

INSTRUMENTS: SP Z helicorder records (USGS)

LOCATION: Tonga Islands

COORDINATES: 20.35 S, 178.75 W

DATE: 30 November 1977

TIME: 10 15 48.4 UTC

DEPTH: 600 km

MAGNITUDE: 5.1 M_S (ALQ)

COMMENTS:

The short-period helicorder records illustrate how the same deep earthquake is recorded at several western U.S. stations. The P arrival at KPK precedes pP by 118 seconds, indicating the depth. All the other pP arrivals are 125 seconds after P, despite increasing distance from the epicenter. The inconsistency on A may be due to conditions at the station.

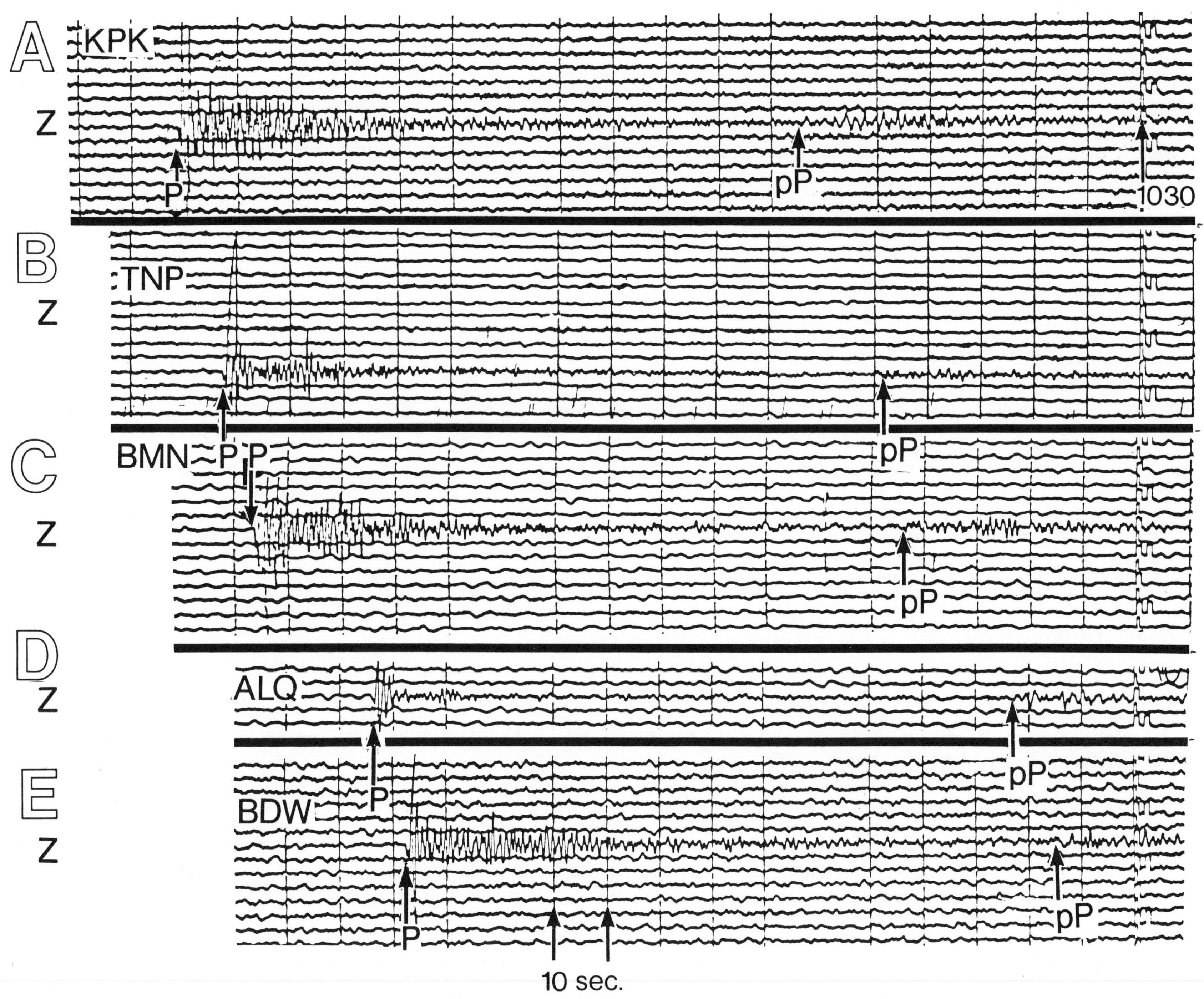
A
KPK
Z
P
pP
1030
B
TNP
Z
pP
C
BMN P P
Z
pP
pP
D
ALQ
Z
pP
E
BDW
Z
P
P
pP
pP
10 sec.

Figure 40

	A	B
DISTANCE:	91.5°	93°
RECORDING STATION:	BLR Black Rapids, Alaska (USGS)	FYU Fort Yukon, Alaska (USGS)
INSTRUMENTS:	SP Z	SP Z
LOCATION:	Offshore Peru	Offshore Peru
COORDINATES:	12.28 S, 76.40 W	12.28 S, 76.40 W
DATE:	5 January 1974	5 January 1974
TIME:	08 33 12.3 UTC	08 33 12.3 UTC
DEPTH:	98 km	98 km
MAGNITUDE:	6.6 M_s (PAS), 6.3 M_b (USGS)	6.6 M_s (PAS), 6.3 M_b (USGS)

COMMENTS:

Ten persons were killed, many injured, and there was extensive property damage in the Lima and Youyos areas (Arequipa).

These records should be compared with those of the deep Tonga Islands earthquake (Figure 39) recorded at USGS stations in western United States.

This intermediate-depth earthquake from Peru recorded at two stations in Alaska indicates the uncertainties of depth determinations. sP is one of the most difficult phases to observe accurately. Different velocity structures at the stations are possibly the cause of discrepancies in the depth phases, pP and sP.

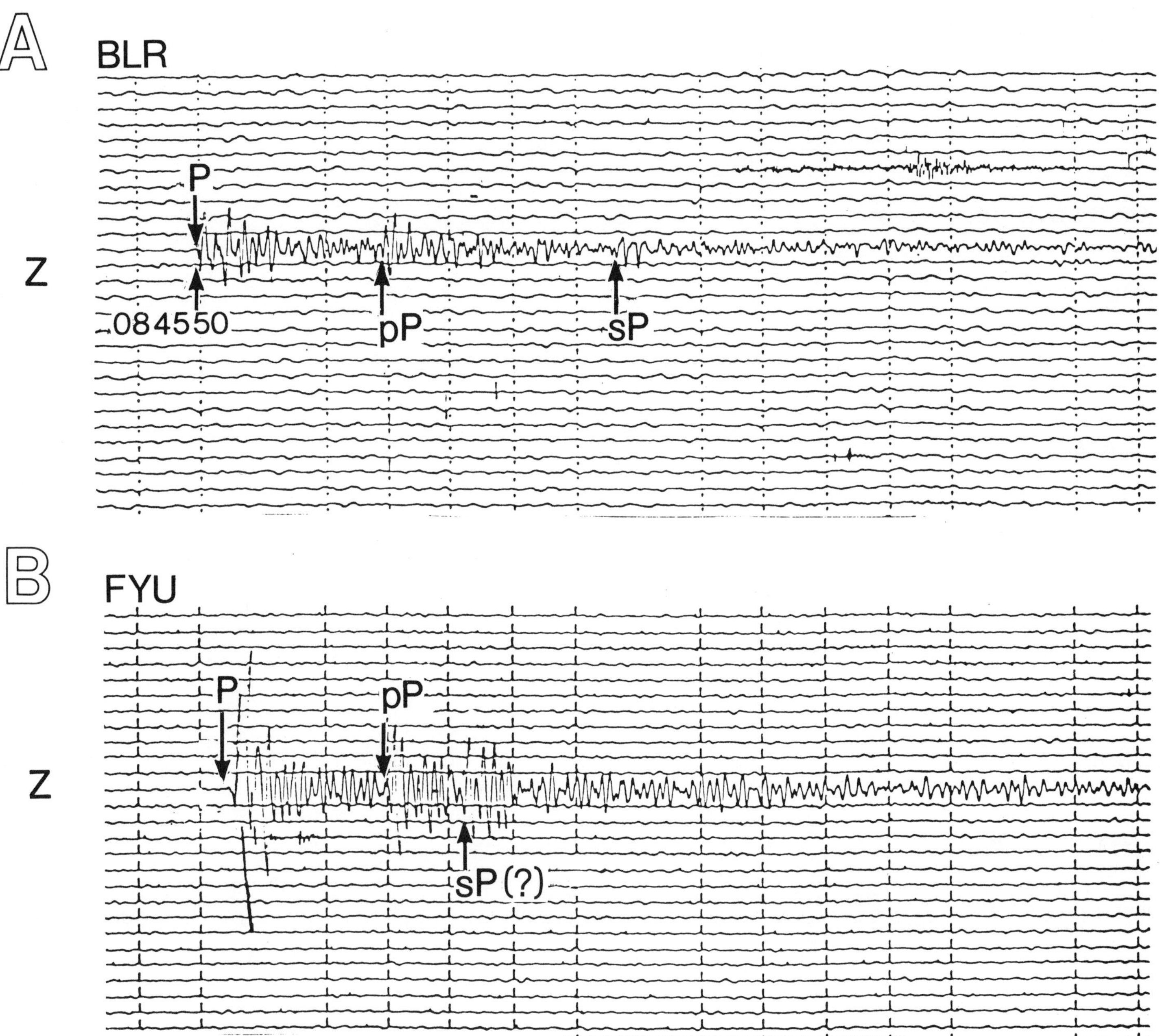

A
BLR
P
Z
084550
pP
sP
B
FYU
P
pP
Z
sP(?)

Figure 41

DISTANCE: 92.2°

RECORDING STATION: GOL

INSTRUMENTS: LP Z, N, E

LOCATION: South of Fiji Islands

COORDINATES: 24.70 S, 177.50 W

DATE: 12 August 1967

TIME: 09 39 44.3 UTC

DEPTH: 134 km

MAGNITUDE: 6.4 M_b (GOL), 6 1/2 M_b (PAS), 6.1 to 6.3 M_b (BRK)

COMMENTS:

In these records, S is not seen. SKS, SKKS, and multiples of S are prominent. Large phases, as SKKS with several branches, are seen with mantle Love and Rayleigh waves in this intermediate-depth earthquake.

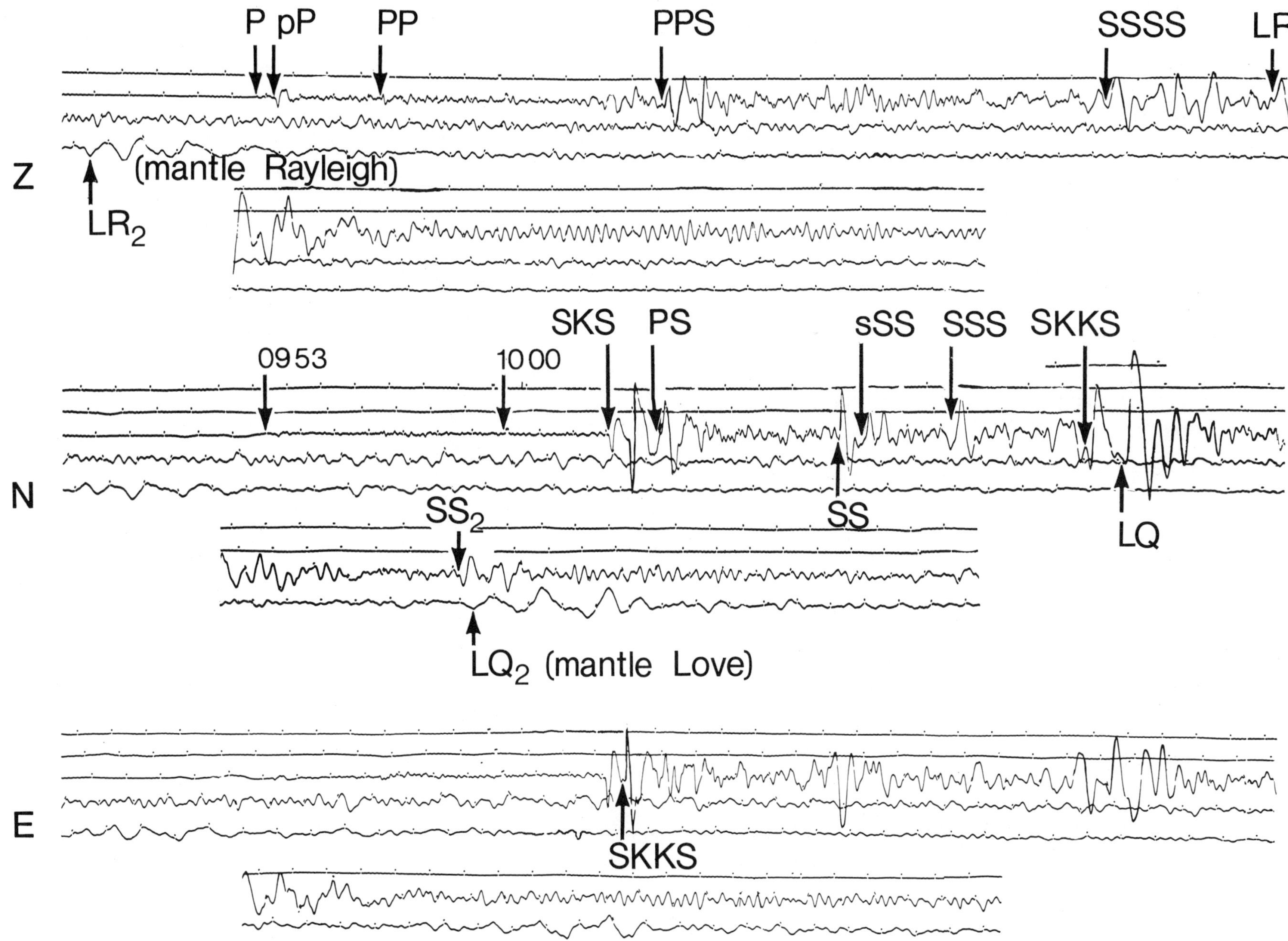

P pP
PP
PPS
SSSS
LR
Z
LR_2
(mantle Rayleigh)
SKS PS
sSS SSS SKKS
0953
1000
N
SS_2
SS
LQ
LQ_2 (mantle Love)
E
SKKS

Figure 42

	A	B
DISTANCE:	99.5°	100°
RECORDING STATION:	GOL	GOL
INSTRUMENTS:	LP Z, N, E	E-W Strainmeter (Filtered)
LOCATION:	Solomon Islands	New Hebrides Islands
COORDINATES:	10.60 S, 161.40 E	15.80 S, 167.2 E
DATE:	13 January 1967	11 August 1965
TIME:	13 48 11.7 UTC	22 31 45.9 UTC
DEPTH:	32 km	130 km
MAGNITUDE:	5.7 M_s (GOL)	6.4 M_b (USGS), 7 1/4 to 7 1/2 M_s (PAS) 7 1/2 to 7 3/4 M_s (BRK)

COMMENTS:

The continuous, sinusoidal train of surface waves characteristic of an oceanic path are noteworthy.

Both branches of mantle Rayleigh wave are seen. Short-period normal dispersion is superimposed on the long-period, inverse dispersion The instrument response is flat out to 600 seconds.

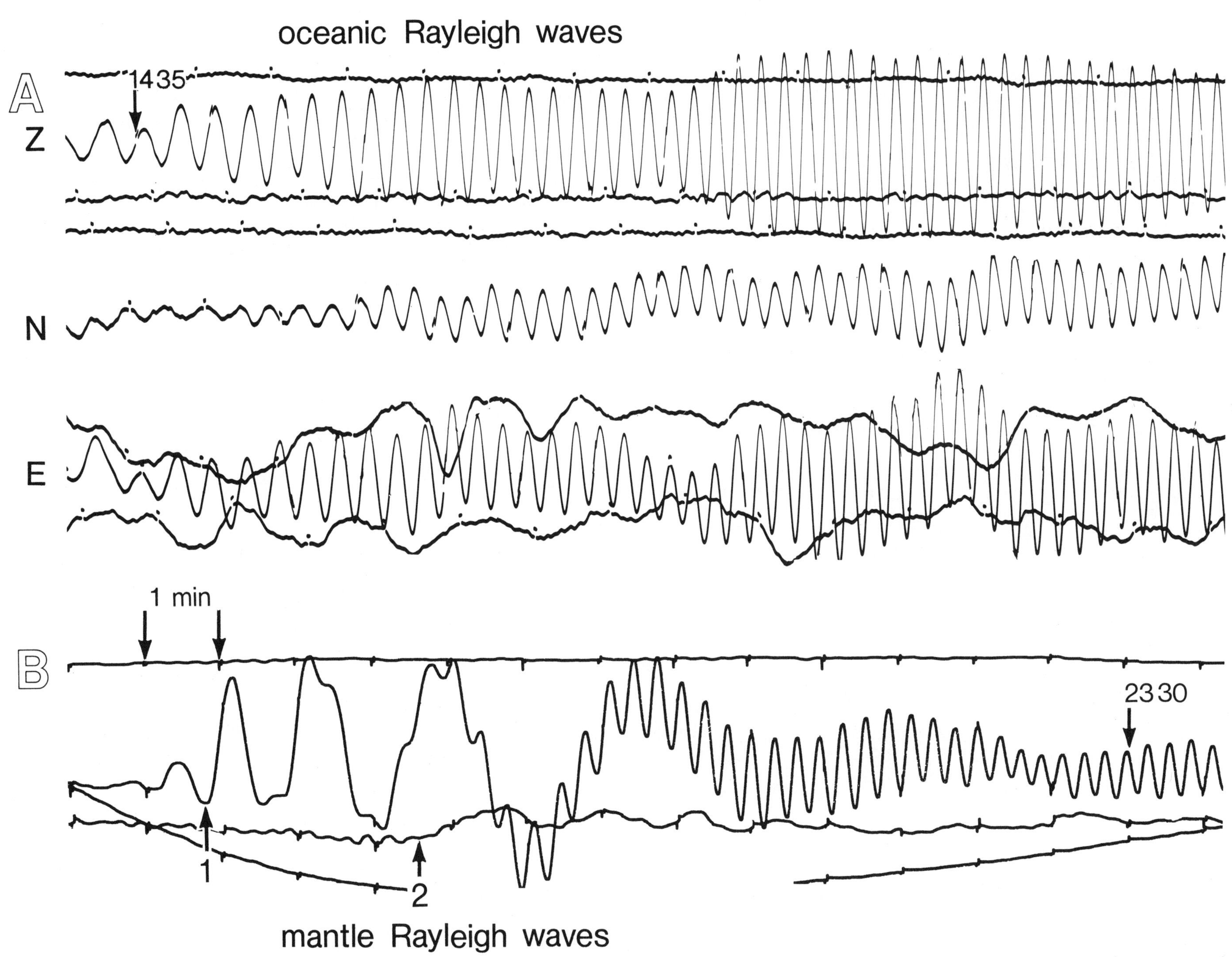

oceanic Rayleigh waves
A
Z
1435
N
E
1 min
B
2330
1
2
mantle Rayleigh waves

Figure 43

DISTANCE: 108°

RECORDING STATION: GOL

INSTRUMENTS: LP Z, N, E

LOCATION: East New Guinea Region

COORDINATES: 7.10 S, 148.30 E

DATE: 23 December 1966

TIME: 15 50 20.4 UTC

DEPTH: 43 km

MAGNITUDE: 6 1/2 to 6 3/4 M_S (GOL), 6 3/4 M_S (PAS), 7.1 to 7.3 M_S (BRK)

COMMENTS:

There are many core phases seen in this earthquake, each of which can be separately identified.

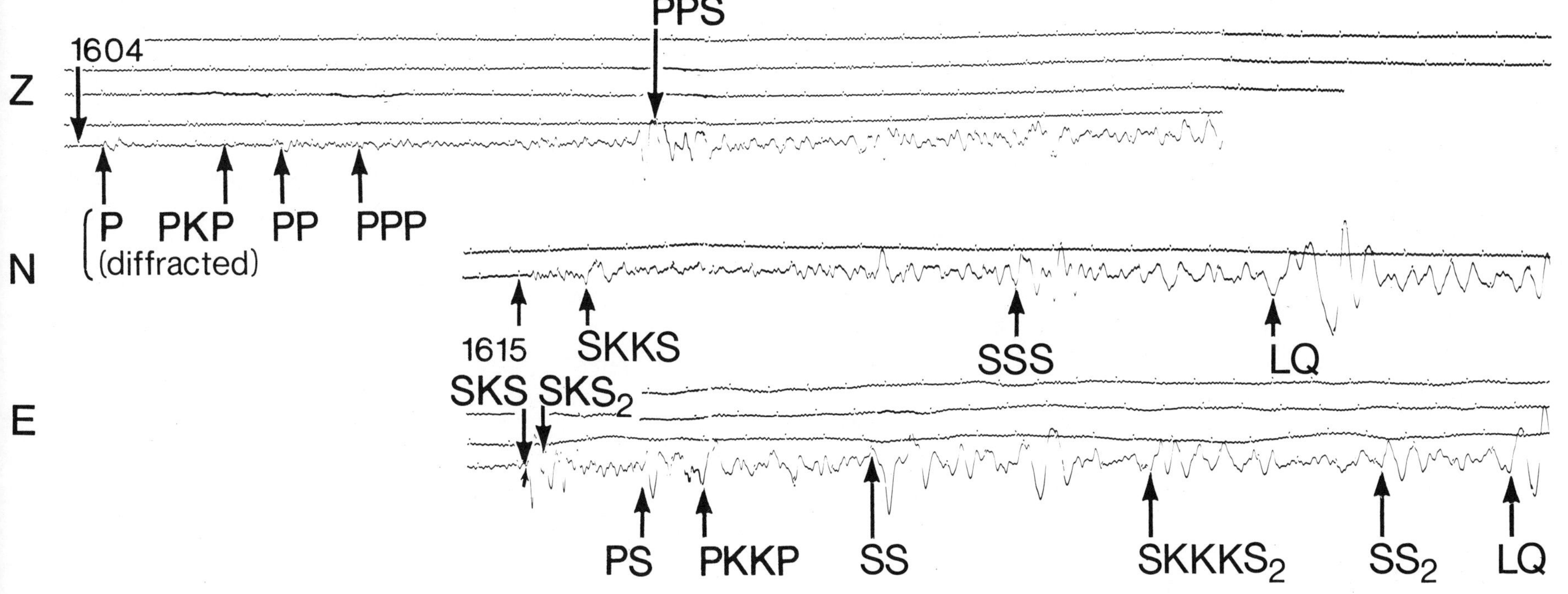

PPS
1604
Z
P PKP PP PPP
(diffracted)
N
1615
SKKS
SSS
LQ
SKS SKS₂
E
SKS SKS2
PS PKKP SS SKKKS2 SS2 LQ

Figure 44

DISTANCE: 122°

RECORDING STATION: GOL

INSTRUMENTS: LP Z, N, E

LOCATION: Bombay, India

COORDINATES: 17.70 N, 73.90 E

DATE: 10 December 1967

TIME: 22 51 24.3 UTC

DEPTH: 33 km

MAGNITUDE: 6.0 M_b (USGS), 6 1/4 to 6 1/2 M_s (GOL), 6 1/2 M_s (PAS), 6.4 to 6.6 M_s (BRK)

COMMENTS:

One hundred persons were killed, 1300 injured in this earthquake, and there was major damage at Kayna Nagar. Felt reports came from a wide area of southwestern India.

The beginning phases are weak; the first phase seen is PP. Large surface waves appear from an earthquake having a mixed path, both oceanic and continental. The second trace of each component is a continuation of the first.

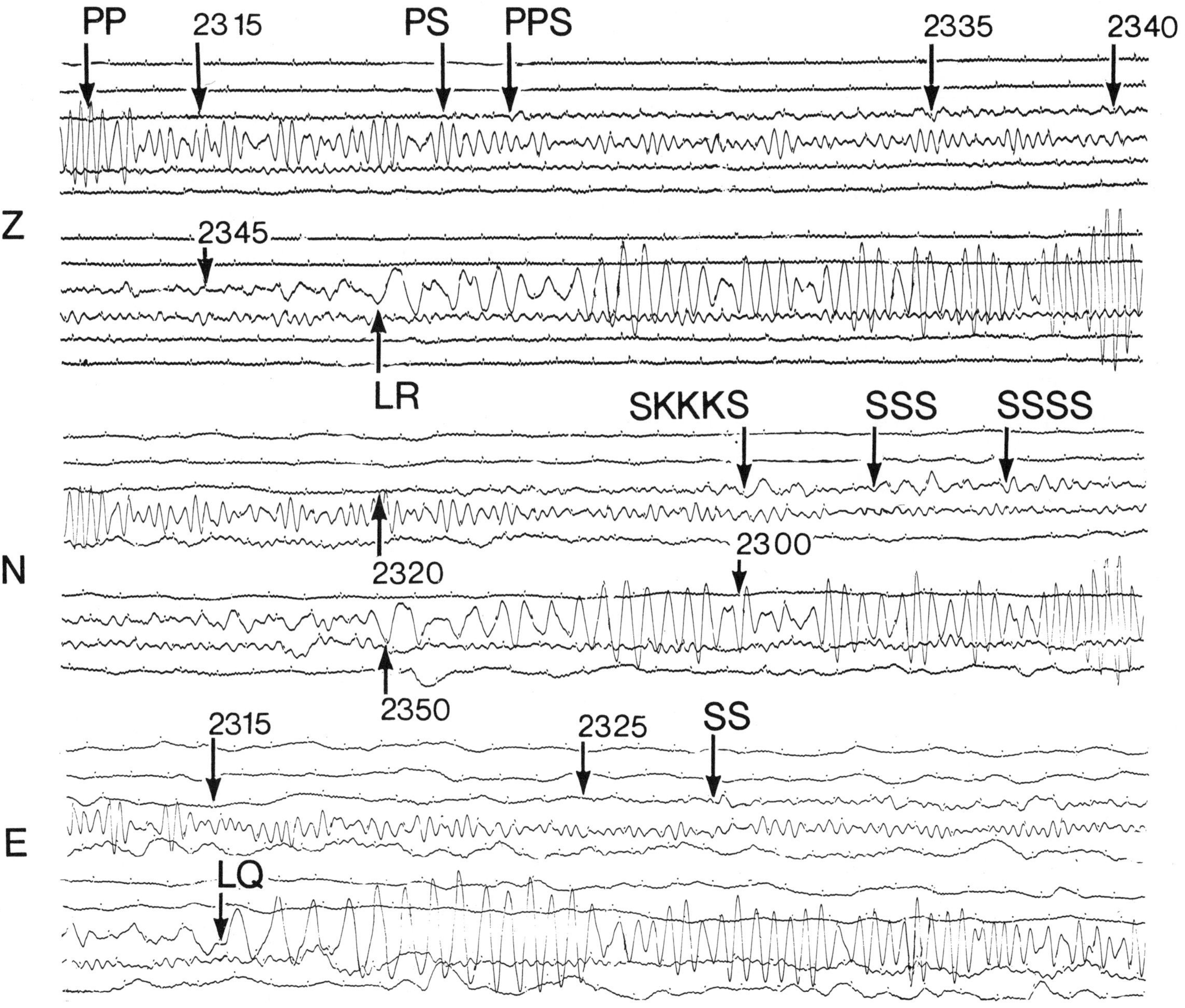

PP
2315
PS
PPS
2335
2340
Z
2345
LR
SKKKS
SSS
SSSS
2320
2300
N
2350
2315
2325
SS
E
LQ
113

Figure 45

DISTANCE:	135°
RECORDING STATION:	GOL
INSTRUMENTS:	SP Z, N, E
LOCATION:	Java
COORDINATES:	9.20 S, 113.10 E
DATE:	19 February 1967
TIME:	22 14 35.3 UTC
DEPTH:	80 km
MAGNITUDE:	6.2 M_b (USGS)

COMMENTS:

Body phases appeared only on short-period seismograms. No long-period waves were seen, although this earthquake was shallower than the one shown on Figure 46. Two branches of PKP and a large PKS are seen. The common error made in interpreting this kind of event is the identification of these phases as P and S from a regional shock. The frequency content on this record, however, is much different from that of regional events.

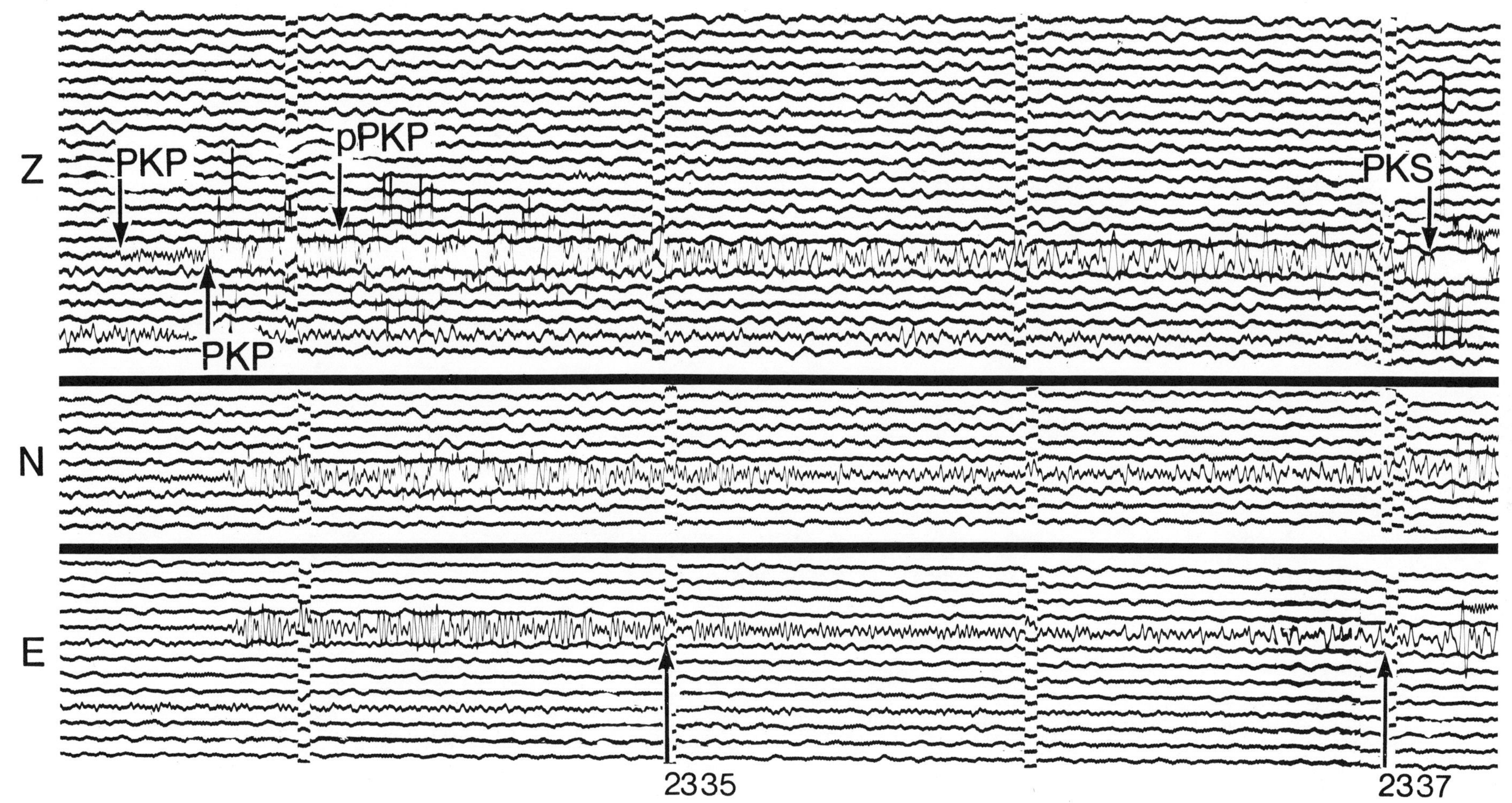

Z
PKP
pPKP
PKP
PKS
N
E
2335
2337

Figure 46

DISTANCE: 139.6°

RECORDING STATION: PAL

INSTRUMENTS: LP Z, N, E

LOCATION: Banda Sea

COORDINATES: 5.70 S, 128.60 E

DATE: 20 August 1965

TIME: 05 54 50.0

DEPTH: 326 km

MAGNITUDE: 6.2 M_b (USGS), 6 3/4 M_s (PAS)

COMMENTS:

This distant earthquake produced long-period body phases not to be confused with surface waves, the beginning of which is difficult to determine in this case, because of the depth of the earthquake.

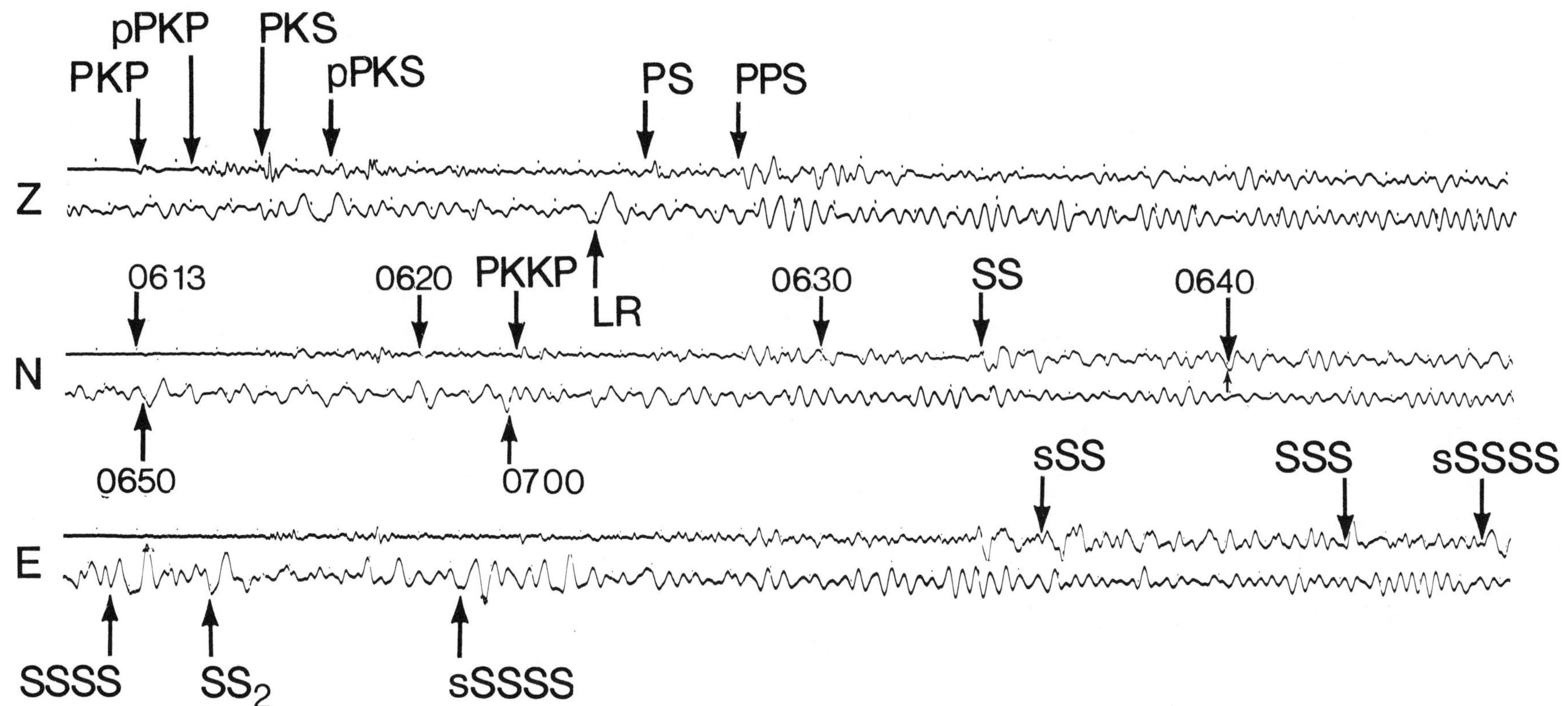
PKP
pPKP
PKS
pPKS
PS
PPS
Z
0613
0620
PKKP
LR
0630
SS
0640
N
0650
0700
sSS
SSS
sSSSS
E
SSSS
SS2
sSSSS

Figure 47

DISTANCE: 700°

RECORDING STATION: GOL

INSTRUMENTS: N-S Strainmeter

LOCATION: Southern Alaska

COORDINATES: 61.00 N, 147.80 W

DATE: 28 March 1964

TIME: 03 36 14.2 UTC

DEPTH: 33 km

MAGNITUDE: 8.5 M_s (GOL)

COMMENTS:

One hundred and fourteen persons were killed or missing in this Great Alaskan Earthquake of Good Friday, 1964. Many were injured and there was major property damage in Alaska. Seismic sea waves caused extensive damage throughout the Gulf of Alaska, the west coast of North America, and Hawaii.

This unique recording is from an unfiltered N-S strainmenter, the response of which falls at about 6 db/octave across the range of frequencies seen. Especially noteworthy are the long-period dispersed Love waves having periods of 10 to 6 minutes and velocities near 5 km/sec. These are considered free oscillations of the earth. The mantle Love wave (G5) has travelled 700° around the earth five times from the epicenter.

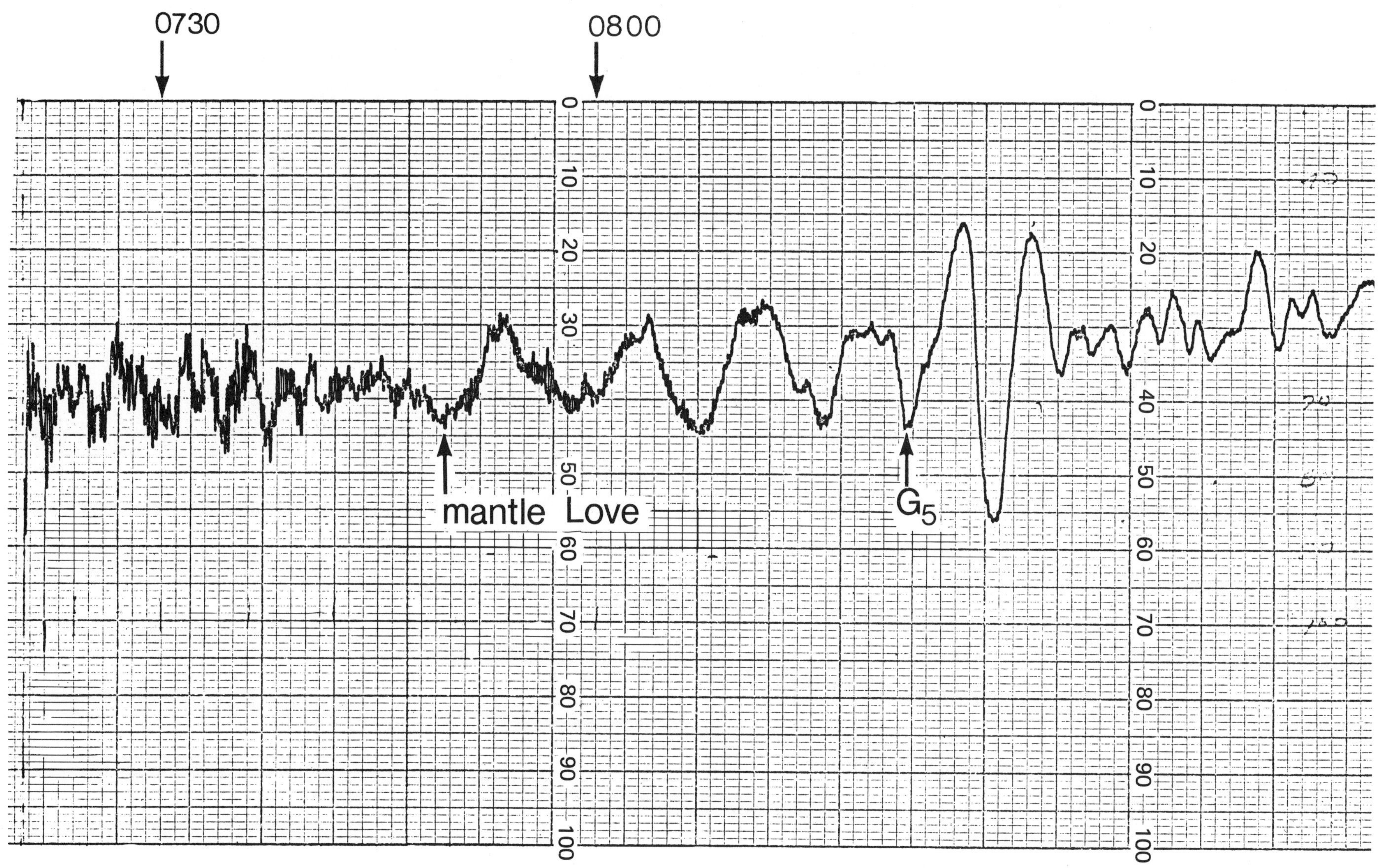

0730
0800
0
10
20
30
40
50
60
70
80
90
100
0
10
20
30
40
50
60
70
80
90
100
mantle Love
G5

Selected Bibliography

The following selected references contain information on some of the specific types of waves marked on the records about which students or researchers might want more information. The list generally reflects early literature; modern references are extensive and are elaborative of these basic papers.

The General Texts and Summaries that follow the list of selected references are books, periodicals, or publications that provide up-to-date sources of general seismological information.

Alsop, L., Sutton, G., and Ewing, M., 1961, Free oscillations of the earth observed on strain and pendulum seismographs, *Journal of Geophysical Research*, v. 66, p. 631-641.

Bolt, B. A., and Dorman, J., 1961, Phase and group velocities of Rayleigh waves in a spherical, gravitating earth, *Journal of Geophysical Research*, v. 66, p. 2965-2981.

Brune, J., Ewing, M., and Kuo, J., 1961, Group and phase velocities for Rayleigh waves of period greater than 380 seconds, *Science*, v. 133, p. 757.

Chander, R., Alsop, L., and Oliver, J., 1968, On the synthesis of shear-coupled PL waves, *Bulletin of the Seismological Society of America*, v. 58, p. 1849-1877.

Ewing, M., 1958, Normal modes of continental surface waves, *Bulletin of the Seismological Society of America*, v. 48, p. 33-49.

Ewing, M., and Press, F., 1950, Crustal structure and surface-wave dispersion, Part I, *Bulletin of the Seismological Society of America*, v. 40, p. 271-280.

Ewing, M., and Press, F., 1952, Crustal structure and surface-wave dispersion, Part II, *Bulletin of the Seismological Society of America*, v. 42, p. 315-325.

Gutenberg, B., and Richter, C. F., 1942, Earthquake magnitude, intensity, energy, and acceleration, *Bulletin of the Seismological Society of America*, v. 32, p. 163; and 1956, v. 46, p. 105.

Jardetzky, W. S., and Press, F., 1953, Crustal structure and surface-wave dispersion, Part III, *Bulletin of the Seismological Society of America*, v. 43, p. 137-144.

Kuo, J., Brune, J., and Major, M., 1962, Rayleigh wave dispersion in the Pacific Ocean for the period range 20 to 140 seconds, *Bulletin of the Seismological Society of America*, v. 52, p. 333-357.

Lehmann, I., 1957, On Lg as read on North American records, *Annali de Geofisica*, v. 10, p. 1-21.

Major, M., Sutton, G., Oliver, J., and Metsger, R., 1964, On elastic strain of the earth in the period range 5 seconds to 100 hours, *Bulletin of the Seismological Society of America*, v. 54, p. 295-346.

Oliver, J., 1962, A summary of observed seismic wave dispersion, *Bulletin of the Seismological Society of America*, v. 52, p. 81-86.

Oliver, J., and Dorman, J., 1961, On the nature of oceanic seismic surface waves with predominant periods of 6 to 8 seconds, *Bulletin of the Seismological Society of America*, v. 51, p. 437-455.

Oliver, J., Dorman, J., and Sutton, G., 1959, The second shear mode of continental Rayleigh waves, *Bulletin of the Seismological Society of America*, v. 49, p. 379-389.

Oliver, J. and Ewing, M., 1957, Microseisms in the 11 to 18 second period range, *Bulletin of the Seismological Society of America*, v. 47, p. 111-127.

Oliver, J., and Ewing, M., 1958, The effect of surficial sedimentary layers on continental surface waves, *Bulletin of the Seismological Society of America*, v. 48, p. 379-389.

Oliver, J., and Ewing, M., 1958, Short-period oceanic surface waves of the Rayleigh and first shear modes, *American Geophysical Union Transactions*, v. 39, p. 482-485.

Oliver, J., Ewing, M., and Press, F., 1955, Crustal studies of the Arctic region from the Lg phase, *Geological Society of America Bulletin*, v. 66, p. 1063-1074.

Oliver, J., Ewing, M, and Press, F., 1955, Crustal structure and surface wave dispersion, Part IV, Atlantic and Pacific Ocean basins, *Geological Society of America Bulletin*, v. 66, p. 913-946.

Oliver, J. and Major, M., 1960, Leaking modes and the P_T phase, *Bulletin of the Seismological Society of America*, v. 50, p. 165-180.

Willmore, P. L., 1957, The use of the stereographic net for the rapid approximate determination of distance and azimuth, *Earthquake Notes*, v. 28, p. 4-6.

General Texts and Summaries

Aki, K., and Richards, P. G., 1980, *Quantitative Seismology*, San Francisco, W. H. Freeman, 2 v.

Båth, M., 1973, *Introduction to Seismology*, New York, John Wiley and Sons, 395 p.

Bolt, B. A., 1978, *Earthquakes, A Primer*, San Francisco, W. H. Freeman, 241 p.

Bullen, K. E., 1963, *An Introduction to the Theory of Seismology*, London, Cambridge University Press, 381 p.

Bulletin of the Seismological Society of America, published bimonthly, Berkeley, California.

Earthquake Engineering Research Institute, *Newsletter*; special publications on significant earthquakes; and field guide for studying earthquakes, *Learning From Earthquakes*, available: 2620 Telegraph Avenue, Berkeley, California 94704.

Gutenberg, B., and Richter, C. F., 1965,
*Seismicity of the Earth and Associated
Phenomena*, New York, Hafner, 310 p.

Jeffreys, H., 1976, *The Earth*, 6th ed., London,
Cambridge University Press, 574 p.

Journal of Geophysical Research, published monthly
by the American Geophysical Union, Washington,
D.C.

Karnik, V., *Seismicity of the European Area*, Part
I, 1969; Part II, 1971, Holland, Dordrecht,
364 p. and 218 p.

Press, F., and Siever, R., 1974, *Earth*, San
Francisco, W. H. Freeman, 945 p.

Richter, C. F., 1958, *Elementary Seismology*, San
Francisco, W. H. Freeman, 768 p.

Rothé, J. P., 1969, *The Seismicity of the Earth
1953-1965*, Paris, UNESCO, 336 p.

U.S. Geological Survey, National Earthquake
Information Service, *Preliminary Determination
of Epicenters*, published weekly; also monthly
listing.

U.S. Geological Survey, National Earthquake
Information Service, *Earthquake Data Reports*,
published biweekly.

U.S. Geological Survey, *Earthquake Information
Bulletin*, published monthly.

Appendix

Seismological Tables

by

Harold Jeffreys and K. E. Bullen

SEISMOLOGICAL TABLES

BY

HAROLD JEFFREYS
M.A., D.Sc., F.R.S.

AND

K. E. BULLEN
M.A., B.Sc., Ph.D.

LONDON

OFFICE OF THE BRITISH ASSOCIATION

BURLINGTON HOUSE, W.1

1967

First Published 1940
Printed by
Neill & Co. Ltd., Edinburgh

Reprinted 1958
by Percy Lund, Humphries
The Country Press, Bradford

Reprinted 1967
by Smith & Ritchie Ltd., Edinburgh

INTRODUCTION

By Harold Jeffreys

A solid can transmit both longitudinal and transverse waves through its interior, the former travelling faster ; hence the longitudinal waves are usually denoted by the letter P for *primary* and the transverse ones by S for *secondary*. The words *push* or *pull* and *shake* may be taken as descriptive of the movement. At any place they have definite velocities independent of the wave-length ; but in the earth the velocities vary with distance from the centre and the paths, corresponding to the rays in optics, are curved. At interfaces between different materials, and at free surfaces, reflexion can take place, and refraction also occurs at interfaces. The behaviour on reflexion and refraction is more complicated than in optics, because in general two new waves in each medium are generated ; thus a P-wave incident at a free surface gives both a reflected P and a reflected S, and so does an incident S, while either type incident on an interface gives a reflected P and S and also a transmitted pair. The angles of incidence are related according to the principle of stationary time in all cases. A P wave reflected at the outer surface as P is denoted by PP, one reflected as S by PS ; similarly an S wave reflected as S is denoted by SS, and one reflected as P by SP. This notation can be extended indefinitely ; thus $SSPP$ would mean a wave reflected three times, being of S type in the first two stages of its path and of P type in the third and fourth. Many earthquakes occur at appreciable depths, and for these there is another complication because reflected waves then do not occur within a certain distance, and beyond that distance they are duplicated. PP in a shallow earthquake, for instance, is reflected at half the distance to the observing station ; in a deep one reflexion occurs at about this distance, and this reflexion is still denoted by PP, but there is also reflexion near the epicentre, analogous to the rays apparently proceeding from the image in a concave mirror. These near reflexions are denoted by pP, sP, sS and so on, and the intervals between the arrival of P and pP, S and sS at a given station give valuable information about the depth of the focus.

At a depth of about 0·45 of the radius there is a discontinuity, originally inferred by R. D. Oldham ; the velocity of P there decreases suddenly with depth, and S below it does not appear to exist. For this and other reasons this central core is generally believed to be liquid. Several notations have been used for waves through it ; here they are denoted by K, following

Sohon. Thus a wave starting as P, transmitted into the core, and again emerging as P, is denoted by PKP; one starting as S, passing through the core, and emerging as S, by SKS, and so on. There are also reflected waves PcP, PcS, ScS, reflected on the outside of the core; the letter c is needed here to distinguish them from PP, PS, SS, reflected at the outer surface. The notation can again be extended; thus $SKKS$ is a wave starting as S, transmitted into the core, once reflected inside the core boundary, and emerging as S.

In the continents there are one or more upper layers, resting on the main layer that extends down to the core. Comparison of waves in these layers in different regions seems to show appreciable differences. Most of the evidence appears to favour an upper layer transmitting waves denoted by Pg and Sg, and an intermediate one transmitting $P*$ and $S*$. P and S without suffixes always refer to waves that have passed beneath these layers. The corresponding results for the ocean floors are not known, but it is suggested that the tables computed for continental foci should be used as a basis of comparison until evidence leads to definite estimates of the differences.

Since the publication of the tables of Dr. Bullen and myself in 1935, and their adoption as a basis of comparison in the International Seismological Summary, some systematic errors have been corrected and much additional observational material has been used. Apart from near earthquake phases, empirical times are now available for P, S, PcP, ScS, PKP and SKS, mostly with standard errors of under a second. These have been used as the basic tables, and all others are calculated from them by means of the principle of stationary time. The phases in the upper layers are not altogether satisfactory, since estimates relevant to crustal structure derived from different earthquakes do not agree so closely as would be expected from their apparent accuracy. The structure finally adopted is based on the pP–P and sS–S intervals in Japanese deep-focus earthquakes, and on Stoneley's studies of the dispersion of surface waves. The upper layer is taken to be 15 ± 3 km. thick, the intermediate one 18 ± 4 km. The velocities in these are taken from European studies.

A main object has been to construct a self-consistent standard of comparison. It is somewhat surprising that differences between travel times in different regions are so small, but there is some evidence that they exist. For P, the times rest mostly on European observations of European and American earthquakes, but there are signs that those for South Pacific earthquakes about $50°$ are about $1^{s}\cdot5$ shorter (the ellipticity terms being taken into account) and those in North American ones about $20°$ about 3^{s} longer. In the Japanese deep shocks there was an unexplained variation between different earthquakes in the times of S, with a mean square value of about $1^{s}\cdot3$, in addition to any attributable to errors of focal depth; this was not shared by SKS. A uniform and consistent standard is necessary for the study of such variations; the tables represent average times, so that no change is possible in the basic phases without making the agreement with observation worse for at least some earthquakes.

P and *S* appear to have loops in their time-curves about 20°. The upper parts of the loops have not been observed directly, but have been somewhat tentatively reconstructed from observations of deep-focus earthquakes. The first arrivals are given in the main *P* and *S* tables, the distances where the two earlier branches intersect being indicated by the letters *d* and *r*. The upper branches are given in separate tables. Focal depths are given as fractions of the adopted radius of the lower layer, and refer to the depth below the top of this layer or the base of the continental layers, supposed to be 33 km. below the outer surface. For *P* and *S* the times are tabulated fully; for other phases the effect of depth is given by allowances to be subtracted from the times for a surface focus. For core phases other than *PKP*, *SKS*, *PcP* and *ScS*, the allowances for depth may be taken as equal to those for these phases at distances where $dt/d\Delta$ is the same. Thus if we want the time of *SKKS* at 150° for depth 0·06*R*, we notice that $dt/d\Delta$ is $5^s\cdot5/1°$, which is matched by *SKS* at 94°; the depth allowance is then 89^s, and the calculated time of *SKKS* will be $30^m\ 18^s - 1^m\ 29^s = 28^m\ 49^s$. For *PKS* the depth allowances will be as for *PcP* with the same $dt/d\Delta$, but for *SKP* they will be as for *ScS*. To avoid this complication it is suggested that the times for a surface focus should be used as the standard unless a definite estimate of focal depth is obtainable, so that *PS* and *SP*, *PKS* and *SKP* and so on can be treated alike.

S up to about 25° is not satisfactorily readable in normal earthquakes, but is very clear in deep ones; the times are based on the latter. In most normal earthquakes *S* seems to be small at these distances, though it can be sometimes recognized by the change of direction of the movement. The nature of the large movements scattered over 30^s or so that are often reported as *S* in these conditions is not yet clear, but they are not *S*; and the *S–P* interval at these distances should not be used to estimate the distance if this is required within 3°. At greater distances *S* is satisfactory except that many earthquakes appear to be multiple, and observers often read the *P* of the first shock and the *S* of the second. This complication has sometimes led to the attribution of "high focus" to normal earthquakes.

Times of basic phases and a few others are given to $0^s\cdot1$; the uncertainties of the total times are larger, but often under 1^s, and differencing is easier if the times are given to $0^s\cdot1$. Phases often read but not very satisfactorily, sometimes on account of indistinct beginnings, are given to 1^s at intervals of 1°. Other phases that have occasionally been read or may be read in the future are given to 1^s at wider intervals.

The tables of *K*, *KIK* and *I* represent times within the core itself and the inner core; they are inferred from the basic observed phases and used in the calculation of others. Figs. 1, 2 and 3 show diagrammatically the forms of the time-curves for various core phases. *K*, *SKS* and *SKKS* are of the form shown in fig. 1, with a single loop. *PKP* resembles fig. 2, the cusp at B being an additional feature; while *PcPPKP* and *PcSPKP* resemble fig. 3, which differs from fig. 2 in the fact that the DF branch passes below the cusp at B. The DE branch of *PKP* corresponds to what was until recently considered to be diffracted movement within the shadow

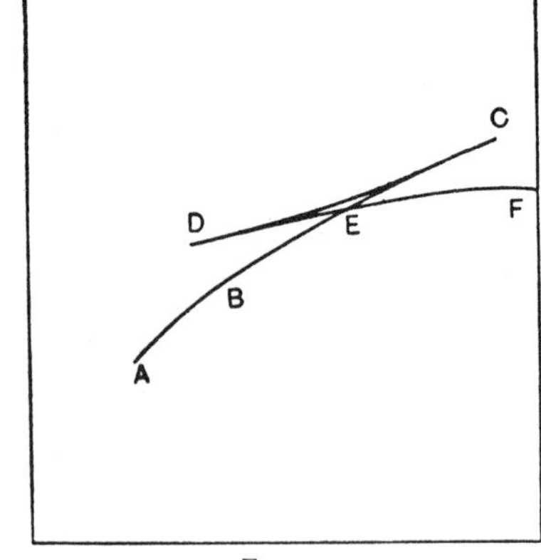

Fig. 1.

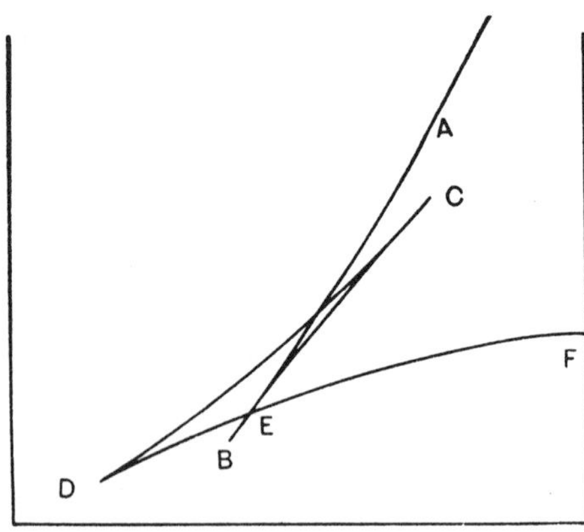

Fig. 2.

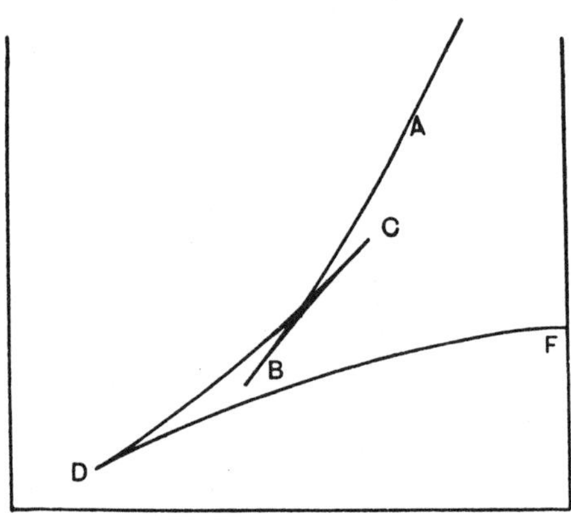

Fig. 3.

zone, but recent work has shown that the nature of the movement is very different from what diffraction would produce, and in these tables the hypothesis of an inner core, introduced by Lehmann and developed by Gutenberg and Richter, has been used. The distribution of velocities corresponding to these times is shown in fig. 4, and the times of the phases, observed or calculated, are shown graphically for a surface focus

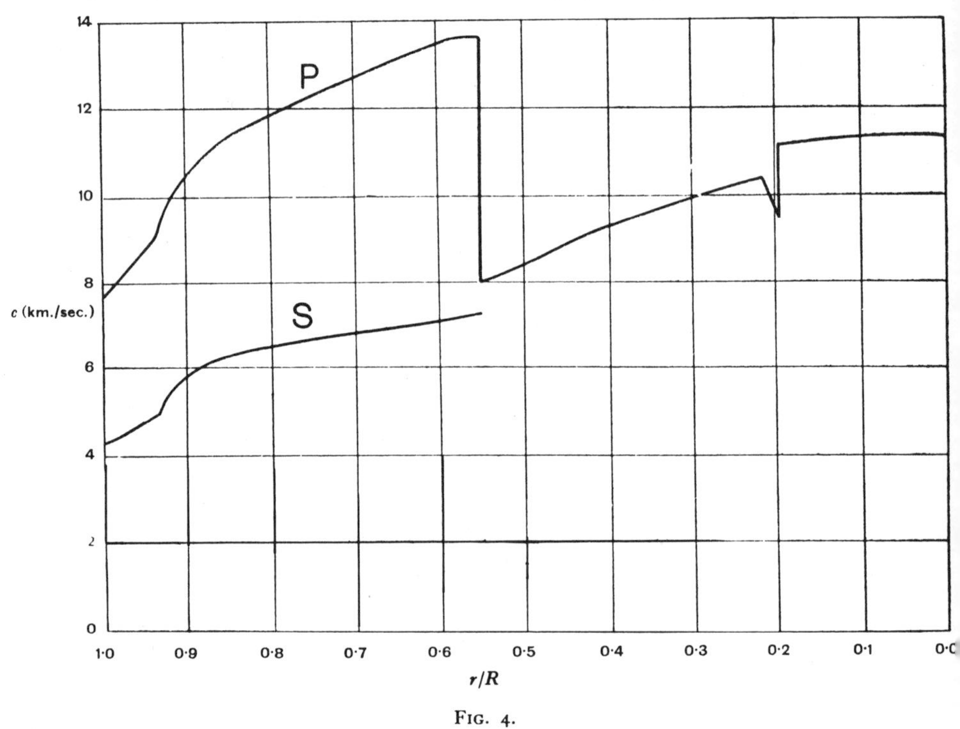

Fig. 4.

in fig. 5. The curves LQ and LR corresponding to the beginnings of the surface waves of Love and Rayleigh types are based on an analysis by Stoneley.

To obtain the highest accuracy geocentric distances should be used. Geocentric direction cosines of the observing stations, compiled by Dr. L. J. Comrie, have been published by the British Association Seismological Committee, which has also published a table due to Bullen of the corrections needed to convert geographic to geocentric distances. Particulars of the method are given in these pamphlets, with references to other literature. Later references, including the construction of the present tables, are as follows :—

BULLEN, K. E. "Theoretical Travel Times of the Phases *PP, PPP, PS, PPS, SS, PSS* and *SSS*," *M.N.R.A.S., Geophys. Suppl.*, **4,** 583–593, 1939.

JEFFREYS, H. "Some Japanese Deep-focus Earthquakes," *M.N.R.A.S.,
Geophys. Suppl.,* **4**, 424–460, 1939.
"The Times of *P*, *S* and *SKS*, and the Velocities of *P* and *S*,"
M.N.R.A.S., Geophys. Suppl., **4**, 498–533, 1939.
"The Times of *PcP* and *ScS*," *M.N.R.A.S., Geophys. Suppl.,* **4**,
537–547, 1939.
"The Times of the Core Waves," *M.N.R.A.S., Geophys. Suppl.,* **4**,
548–561, 1939.
"Times of Transmission for Small Distances and Focal Depths,"
M.N.R.A.S., Geophys. Suppl., **4**, 571–578, 1939.
"The Times of the Core Waves (Second Paper)," *M.N.R.A.S.
Geophys. Suppl.,* **4**, 594–615, 1939.
STONELEY, R. "On the *L* Phase of Seismograms," *M.N.R.A.S.,
Geophys. Suppl.,* **4**, 562–569, 1939.

A description of the construction of the tables is given by

JEFFREYS, H. "Seismological Tables," *M.N.R.A.S.,* **99**, 397–408, 1939.

For general accounts, see

JEFFREYS, H. *The Earth*, Camb. Univ. Press, 1929.
Earthquakes and Mountains, Methuen, 1935.
MACELWANE, J. B., and SOHON, F. W. *Theoretical Seismology*, John
Wiley & Sons, 1932, 1936.
Bulletins of U.S. Nat. Res. Council, "Seismology," 1933, "Internal
Constitution of the Earth," 1939.

The present tables have all been published by the Royal Astronomical Society, and thanks are due to that Society for their kind permission to reproduce them in a collected form.

Δ	Surface	0.00	0.01	0.02	0.03	0.04	0.05	0.06	0.07	0.08	0.09	0.10	0.11	0.12
						Depth $h =$								
0.0	(6.8)	5.4	13.5	21.4	29.1	36.6	43.9	51.1	58.0	1 04.5	1 10.8	1 16.8	1 22.8	1 28.8
0.5	14.0	10.5	15.6	22.6	29.9	37.2	44.4	51.5	58.3	1 04.8	1 11.0	1 17.0	1 23.0	1 29.0
1.0	21.1	17.7	20.4	25.8	32.3	39.1	45.9	52.7	59.3	1 05.7	1 11.8	1 17.7	1 23.6	1 29.5
1.5	28.2	24.8	26.7	30.6	36.0	42.3	48.4	54.8	1 01.0	1 07.2	1 13.1	1 18.8	1 24.6	1 30.3
2.0	35.4	32.0	32.9	36.0	40.6	45.8	51.6	57.6	1 03.4	1 09.2	1 14.8	1 20.3	1 25.9	1 31.5
2.5	42.6	39.1	39.8	42.1	45.7	50.3	55.4	1 00.9	1 06.3	1 11.8	1 17.1	1 22.3	1 27.7	1 33.0
3.0	49.7	46.3	46.7	48.4	51.3	55.2	59.7	1 04.6	1 09.7	1 14.8	1 19.8	1 24.7	1 29.8	1 34.9
3.5	56.8	53.4	53.6	54.8	57.1	1 00.4	1 04.4	1 08.9	1 13.4	1 18.1	1 22.8	1 27.5	1 32.3	1 37.2
4.0	1 03.9	1 00.5	1 00.4	1 01.3	1 03.2	1 06.0	1 09.4	1 13.4	1 17.5	1 21.7	1 26.2	1 30.6	1 35.1	1 39.7
4.5	1 11.0	1 07.6	1 07.2	1 07.8	1 09.4	1 11.7	1 14.7	1 18.2	1 21.9	1 25.7	1 29.7	1 33.7	1 38.3	1 42.4
5.0	1 18.1	1 14.7	1 14.1	1 14.4	1 15.6	1 17.5	1 20.1	1 23.3	1 26.5	1 29.9	1 33.5	1 37.3	1 41.4	1 45.4
5.5	1 25.2	1 21.7	1 20.9	1 21.0	1 22.0	1 23.5	1 25.7	1 28.4	1 31.3	1 34.3	1 37.6	1 40.9	1 44.8	1 48.6
6.0	1 32.2	1 28.7	1 27.7	1 27.6	1 28.3	1 29.5	1 31.4	1 33.7	1 36.3	1 38.9	1 41.8	1 44.8	1 48.2	1 51.9
6.5	1 39.3	1 35.8	1 34.6	1 34.3	1 34.7	1 35.6	1 37.2	1 39.2	1 41.4	1 43.6	1 46.1	1 48.9	1 52.0	1 55.3
7.0	1 46.3	1 42.8	1 41.6	1 41.0	1 41.0	1 41.8	1 43.1	1 44.8	1 46.5	1 48.4	1 50.6	1 53.1	1 55.9	1 58.9
7.5	1 53.3	1 49.8	1 48.4	1 47.7	1 47.5	1 48.0	1 49.0	1 50.4	1 51.8	1 53.3	1 55.2	1 57.4	1 59.9	2 02.6
8.0	2 00.3	1 56.7	1 55.2	1 54.3	1 54.0	1 54.2	1 54.9	1 56.1	1 57.1	1 58.3	1 59.8	2 01.7	2 04.0	2 06.5
8.5	2 07.3	2 03.7	2 02.0	2 00.9	2 00.5	2 00.4	2 00.8	2 01.8	2 02.5	2 03.3	2 04.5	2 06.2	2 08.2	2 10.5
9.0	2 14.2	2 10.6	2 08.7	2 07.5	2 06.9	2 06.6	2 06.8	2 07.5	2 07.9	2 08.4	2 09.3	2 10.7	2 12.5	2 14.5
9.5	2 21.1	2 17.5	2 15.5	2 14.1	2 13.3	2 12.8	2 12.8	2 13.2	2 13.3	2 13.5	2 14.1	2 15.3	2 16.8	2 18.6
10.0	2 28.0	2 24.4	2 22.2	2 20.6	2 19.7	2 19.0	2 18.8	2 19.0	2 18.6	2 18.6	2 18.9	2 19.9	2 21.2	2 22.7
11.0	2 41.7	2 38.1	2 35.6	2 33.7	2 32.5	2 31.4	2 30.8	d2 30.6	2 29.3	2 28.8	2 28.7	2 29.1	2 30.0	2 31.2
12.0	2 55.3	2 51.6	2 48.9	2 46.7	2 45.1	2 43.8	2 42.8	r2 41.8	2 39.9	2 38.9	2 38.6	2 38.6	2 39.0	2 39.8
13.0	3 08.7	3 05.0	3 02.1	2 59.6	2 57.6	2 56.0	d2 54.7	2 52.9	2 50.5	2 49.1	2 48.5	2 48.1	2 48.1	2 48.5
14.0	3 21.9	3 18.1	3 15.0	3 12.3	3 10.0	3 08.1	r3 06.3	3 03.6	3 00.9	2 59.2	2 58.2	2 57.6	2 57.2	2 57.3
15.0	3 35.0	3 31.2	3 27.9	3 24.9	3 22.3	d3 20.1	3 17.0	3 14.0	3 11.1	3 09.2	3 07.9	3 07.1	3 06.4	3 06.2
16.0	3 48.0	3 44.1	3 40.5	3 37.3	d3 34.4	r3 31.3	3 27.6	3 24.1	3 21.2	3 19.0	3 17.5	3 16.4	3 15.5	3 15.1
17.0	4 00.7	3 56.7	3 52.9	d3 49.5	r3 46.0	3 41.9	3 38.0	3 34.2	3 31.2	3 28.7	3 26.9	3 25.5	3 24.5	3 24.0
18.0	4 13.2	4 09.2	d4 05.2	r4 01.1	3 56.5	3 52.2	3 48.1	3 44.3	3 41.1	3 38.4	3 36.3	3 34.7	3 33.5	3 32.8
19.0	d4 25.5	d4 21.5	r4 16.5	4 11.6	4 06.7	4 02.3	3 58.1	3 54.1	3 50.8	3 47.9	3 45.6	3 43.8	3 42.5	3 41.6
20.0	r4 37.0	r4 32.5	4 27.1	4 21.9	4 16.9	4 12.2	4 07.9	4 03.8	4 00.3	3 57.3	3 54.8	3 52.8	3 51.3	3 50.3
21.0	4 47.4	4 42.9	4 37.3	4 31.9	4 26.8	4 22.0	4 17.6	4 13.4	4 09.7	4 06.5	4 03.9	4 01.8	4 00.2	3 59.0
22.0	4 57.5	4 52.9	4 47.2	4 41.7	4 36.5	4 31.7	4 27.2	4 22.9	4 19.0	4 15.7	4 13.0	4 10.8	4 09.0	4 07.7
23.0	5 07.4	5 02.8	4 56.9	4 51.4	4 46.2	4 41.2	4 36.5	4 32.0	4 28.1	4 24.7	4 21.9	4 19.6	4 17.7	4 16.3
24.0	5 17.1	5 12.5	5 06.5	5 01.0	4 55.6	4 50.5	4 45.7	4 41.1	4 37.1	4 33.7	4 30.8	4 28.4	4 26.4	4 24.9
25.0	5 26.8	5 22.2	5 16.2	5 10.4	5 04.9	4 59.6	4 54.7	4 50.1	4 46.1	4 42.6	4 39.6	4 37.1	4 35.0	4 33.4
26.0	5 36.2	5 31.6	5 25.5	5 19.7	5 14.1	5 08.7	5 03.8	4 59.1	4 55.0	4 51.4	4 48.4	4 45.8	4 43.6	4 41.9
27.0	5 45.4	5 40.8	5 34.7	5 28.8	5 23.2	5 17.8	5 12.8	5 08.0	5 03.9	5 00.2	4 57.1	4 54.4	4 52.1	4 50.3
28.0	5 54.5	5 49.9	5 43.8	5 37.9	5 32.2	5 26.8	5 21.7	5 16.8	5 12.6	5 09.0	5 05.7	5 02.9	5 00.6	4 58.7
29.0	6 03.5	5 58.8	5 52.7	5 46.8	5 41.1	5 35.6	5 30.5	5 25.5	5 21.4	5 17.6	5 14.3	5 11.4	5 09.0	5 07.1
30.0	6 12.5	6 07.7	6 01.6	5 55.7	5 49.9	5 44.4	5 39.2	5 34.2	5 30.0	5 26.2	5 22.8	5 19.9	5 17.4	5 15.4

TIMES OF P—*continued*

Depth $h =$

Δ	Surface	0·00	0·01	0·02	0·03	0·04	0·05
30	6 12·5 (88)	6 07·7 (89)	6 01·6 (88)	5 55·7 (88)	5 49·9 (87)	5 44·4 (87)	5 39·2 (86)
31	6 21·3 (88)	6 16·6 (88)	6 10·4 (88)	6 04·5 (87)	5 58·6 (87)	5 53·1 (87)	5 47·8 (86)
32	6 30·1 (87)	6 25·4 (87)	6 19·1 (87)	6 13·2 (87)	6 07·3 (87)	6 01·7 (86)	5 56·4 (86)
33	6 38·8 (87)	6 34·1 (86)	6 27·8 (87)	6 21·8 (86)	6 15·9 (85)	6 10·3 (85)	6 04·9 (85)
34	6 47·5 (86)	6 42·7 (86)	6 36·4 (85)	6 30·3 (85)	6 24·4 (85)	6 18·8 (84)	6 13·4 (84)
35	6 56·1 (85)	6 51·3 (85)	6 44·9 (85)	6 38·8 (84)	6 32·9 (84)	6 27·2 (84)	6 21·8 (84)
36	7 04·6 (84)	6 59·8 (84)	6 53·4 (84)	6 47·2 (84)	6 41·3 (84)	6 35·6 (84)	6 30·2 (83)
37	7 13·0 (84)	7 08·2 (84)	7 01·8 (83)	6 55·6 (84)	6 49·7 (83)	6 44·0 (83)	6 38·5 (83)
38	7 21·4 (84)	7 16·6 (83)	7 10·1 (83)	7 04·0 (83)	6 58·0 (83)	6 52·3 (82)	6 46·8 (82)
39	7 29·8 (83)	7 24·9 (83)	7 18·4 (83)	7 12·3 (82)	7 06·3 (82)	7 00·5 (82)	6 55·0 (82)
40	7 38·1 (82)	7 33·2 (83)	7 26·7 (82)	7 20·5 (82)	7 14·5 (82)	7 08·7 (82)	7 03·2 (81)
41	7 46·3 (82)	7 41·5 (82)	7 34·9 (82)	7 28·7 (82)	7 22·7 (82)	7 16·9 (81)	7 11·3 (81)
42	7 54·5 (82)	7 49·7 (82)	7 43·1 (82)	7 36·9 (81)	7 30·8 (81)	7 25·0 (81)	7 19·4 (80)
43	8 02·7 (81)	7 57·9 (81)	7 51·3 (81)	7 45·0 (81)	7 38·9 (81)	7 33·0 (80)	7 27·4 (80)
44	8 10·8 (81)	8 06·0 (80)	7 59·4 (80)	7 53·0 (80)	7 46·9 (79)	7 41·0 (79)	7 35·4 (79)
45	8 18·9 (79)	8 14·0 (80)	8 07·4 (79)	8 01·0 (79)	7 54·8 (79)	7 48·9 (78)	7 43·3 (78)
46	8 26·8 (79)	8 22·0 (78)	8 15·3 (79)	8 08·9 (79)	8 02·7 (77)	7 56·7 (78)	7 51·1 (77)
47	8 34·7 (79)	8 29·8 (79)	8 23·2 (78)	8 16·7 (78)	8 10·4 (78)	8 04·5 (78)	7 58·8 (77)
48	8 42·6 (77)	8 37·7 (77)	8 31·0 (77)	8 24·5 (78)	8 18·2 (76)	8 12·2 (77)	8 06·5 (76)
49	8 50·3 (77)	8 45·4 (77)	8 38·7 (77)	8 32·2 (77)	8 25·8 (76)	8 19·8 (76)	8 14·1 (75)
50	8 58·0 (76)	8 53·1 (76)	8 46·4 (76)	8 39·8 (76)	8 33·4 (75)	8 27·3 (74)	8 21·6 (74)
51	9 05·6 (76)	9 00·7 (75)	8 54·0 (75)	8 47·4 (75)	8 40·9 (74)	8 34·7 (74)	8 29·0 (74)
52	9 13·2 (75)	9 08·2 (75)	9 01·5 (75)	8 54·9 (74)	8 48·3 (74)	8 42·1 (73)	8 36·4 (73)
53	9 20·7 (73)	9 15·7 (74)	9 09·0 (73)	9 02·3 (73)	8 55·7 (73)	8 49·4 (72)	8 43·7 (72)
54	9 28·0 (74)	9 23·1 (73)	9 16·3 (73)	9 09·6 (73)	9 03·0 (72)	8 56·6 (72)	8 50·9 (71)
55	9 35·4 (72)	9 30·4 (72)	9 23·6 (72)	9 16·9 (72)	9 10·2 (71)	9 03·8 (71)	8 58·0 (71)
56	9 42·6 (72)	9 37·6 (72)	9 30·8 (72)	9 24·1 (71)	9 17·3 (71)	9 10·9 (71)	9 05·1 (71)
57	9 49·8 (72)	9 44·8 (72)	9 37·9 (71)	9 31·1 (70)	9 24·4 (70)	9 17·9 (70)	9 12·0 (69)
58	9 56·8 (70)	9 51·8 (70)	9 44·9 (70)	9 38·1 (70)	9 31·4 (70)	9 24·9 (69)	9 18·9 (69)
59	10 03·8 (69)	9 58·8 (69)	9 51·8 (69)	9 45·1 (68)	9 38·3 (68)	9 31·8 (69)	9 25·8 (67)
60	10 10·7 (68)	10 05·7 (68)	9 58·7 (68)	9 51·9 (68)	9 45·1 (67)	9 38·6 (67)	9 32·5 (66)
61	10 17·5 (68)	10 12·5 (67)	10 05·5 (67)	9 58·7 (67)	9 51·8 (67)	9 45·3 (66)	9 39·1 (66)
62	10 24·3 (66)	10 19·2 (67)	10 12·2 (67)	10 05·4 (66)	9 58·5 (66)	9 51·9 (65)	9 45·7 (65)
63	10 30·9 (66)	10 25·9 (66)	10 18·9 (65)	10 12·0 (65)	10 05·1 (65)	9 58·4 (65)	9 52·2 (64)
64	10 37·5 (65)	10 32·4 (65)	10 25·4 (65)	10 18·5 (65)	10 11·6 (64)	10 04·9 (64)	9 58·6 (64)
65	10 44·0 (64)	10 38·9 (64)	10 31·9 (64)	10 25·0 (64)	10 18·0 (64)	10 11·3 (63)	10 05·0 (63)
66	10 50·4 (64)	10 45·3 (64)	10 38·3 (64)	10 31·4 (63)	10 24·4 (63)	10 17·6 (63)	10 11·3 (63)
67	10 56·8 (63)	10 51·6 (63)	10 44·6 (62)	10 37·7 (62)	10 30·7 (62)	10 23·9 (62)	10 17·6 (61)
68	11 03·1 (62)	10 57·9 (62)	10 50·8 (62)	10 43·9 (61)	10 36·9 (61)	10 30·1 (62)	10 23·7 (61)
69	11 09·3 (61)	11 04·1 (61)	10 57·0 (61)	10 50·0 (61)	10 43·0 (61)	10 36·2 (61)	10 29·8 (61)
70	11 15·4	11 10·2	11 03·1	10 56·1	10 49·1	10 42·3	10 35·9

TIMES OF P—*continued*

Depth $h =$

Δ	0·06	0·07	0·08	0·09	0·10	0·11	0·12
30	5 34·2 (86)	5 30·0 (86)	5 26·2 (85)	5 22·8 (85)	5 19·9 (84)	5 17·4 (83)	5 15·4 (83)
31	5 42·8 (86)	5 38·6 (85)	5 34·7 (85)	5 31·3 (85)	5 28·3 (84)	5 25·7 (84)	5 23·7 (82)
32	5 51·4 (86)	5 47·1 (84)	5 43·2 (84)	5 39·7 (84)	5 36·7 (84)	5 34·1 (84)	5 31·9 (82)
33	5 59·9 (85)	5 55·5 (84)	5 51·6 (84)	5 48·1 (84)	5 45·0 (83)	5 42·3 (82)	5 40·1 (81)
34	6 08·4 (84)	6 03·9 (84)	6 00·0 (83)	5 56·4 (83)	5 53·3 (82)	5 50·5 (82)	5 48·2 (81)
35	6 16·8 (84)	6 12·3 (83)	6 08·3 (83)	6 04·7 (82)	6 01·5 (82)	5 58·7 (81)	5 56·3 (81)
36	6 25·2 (83)	6 20·6 (83)	6 16·6 (82)	6 12·9 (82)	6 09·7 (82)	6 06·8 (81)	6 04·4 (80)
37	6 33·5 (82)	6 28·9 (82)	6 24·8 (81)	6 21·1 (82)	6 17·9 (81)	6 14·9 (81)	6 12·4 (80)
38	6 41·7 (82)	6 37·1 (82)	6 32·9 (81)	6 29·3 (81)	6 26·0 (80)	6 23·0 (80)	6 20·4 (79)
39	6 49·9 (82)	6 45·3 (81)	6 41·0 (81)	6 37·4 (80)	6 34·0 (80)	6 31·0 (79)	6 28·3 (78)
40	6 58·1 (81)	6 53·4 (81)	6 49·1 (80)	6 45·4 (80)	6 42·0 (79)	6 38·9 (78)	6 36·1 (77)
41	7 06·2 (80)	7 01·5 (80)	6 57·1 (80)	6 53·4 (79)	6 49·9 (79)	6 46·7 (78)	6 43·8 (77)
42	7 14·2 (80)	7 09·5 (80)	7 05·1 (79)	7 01·3 (79)	6 57·8 (79)	6 54·5 (78)	6 51·5 (77)
43	7 22·2 (80)	7 17·4 (79)	7 13·0 (79)	7 09·1 (78)	7 05·5 (78)	7 02·2 (77)	6 59·1 (76)
44	7 30·1 (79)	7 25·3 (78)	7 20·8 (78)	7 16·9 (77)	7 13·3 (76)	7 09·8 (76)	7 06·7 (75)
45	7 38·0 (78)	7 33·1 (77)	7 28·6 (77)	7 24·6 (76)	7 20·9 (76)	7 17·4 (75)	7 14·2 (75)
46	7 45·8 (77)	7 40·8 (77)	7 36·3 (76)	7 32·2 (76)	7 28·5 (76)	7 24·9 (75)	7 21·7 (75)
47	7 53·5 (76)	7 48·5 (77)	7 43·9 (76)	7 39·8 (75)	7 36·0 (75)	7 32·4 (75)	7 29·1 (74)
48	8 01·1 (76)	7 56·1 (76)	7 51·5 (75)	7 47·3 (75)	7 43·4 (74)	7 39·8 (74)	7 36·4 (73)
49	8 08·7 (75)	8 03·6 (75)	7 59·0 (75)	7 54·8 (75)	7 50·8 (74)	7 47·1 (73)	7 43·7 (72)
50	8 16·2 (74)	8 11·1 (74)	8 06·4 (73)	8 02·1 (73)	7 58·1 (72)	7 54·4 (72)	7 50·9 (71)
51	8 23·6 (74)	8 18·5 (73)	8 13·7 (73)	8 09·4 (71)	8 05·3 (71)	8 01·6 (70)	7 58·0 (70)
52	8 31·0 (73)	8 25·8 (72)	8 21·0 (71)	8 16·5 (71)	8 12·4 (71)	8 08·6 (70)	8 05·0 (69)
53	8 38·3 (72)	8 33·0 (71)	8 28·1 (71)	8 23·6 (70)	8 19·5 (70)	8 15·6 (70)	8 11·9 (69)
54	8 45·5 (71)	8 40·1 (71)	8 35·2 (70)	8 30·6 (70)	8 26·5 (69)	8 22·6 (68)	8 18·8 (68)
55	8 52·6 (70)	8 47·2 (70)	8 42·2 (69)	8 37·6 (69)	8 33·4 (68)	8 29·4 (68)	8 25·6 (67)
56	8 59·6 (70)	8 54·2 (70)	8 49·1 (69)	8 44·5 (69)	8 40·2 (68)	8 36·2 (67)	8 32·3 (67)
57	9 06·5 (69)	9 01·0 (68)	8 56·0 (68)	8 51·3 (67)	8 47·0 (66)	8 42·9 (66)	8 39·0 (65)
58	9 13·3 (67)	9 07·8 (68)	9 02·8 (67)	8 58·1 (67)	8 53·7 (66)	8 49·5 (66)	8 45·6 (65)
59	9 20·0 (67)	9 14·5 (67)	9 09·5 (66)	9 04·8 (66)	9 00·3 (66)	8 56·1 (65)	8 52·1 (65)
60	9 26·7 (66)	9 21·2 (66)	9 16·1 (66)	9 11·4 (65)	9 06·9 (65)	9 02·6 (64)	8 58·6 (64)
61	9 33·3 (66)	9 27·8 (65)	9 22·7 (65)	9 17·9 (65)	9 13·4 (64)	9 09·0 (64)	9 05·0 (64)
62	9 39·9 (65)	9 34·4 (65)	9 29·2 (64)	9 24·4 (64)	9 19·8 (64)	9 15·4 (64)	9 11·4 (62)
63	9 46·4 (64)	9 40·9 (64)	9 35·6 (64)	9 30·8 (63)	9 26·2 (63)	9 21·8 (62)	9 17·6 (63)
64	9 52·8 (64)	9 47·3 (63)	9 42·0 (63)	9 37·1 (63)	9 32·5 (62)	9 28·0 (62)	9 23·9 (61)
65	9 59·2 (63)	9 53·6 (62)	9 48·3 (62)	9 43·4 (62)	9 38·7 (61)	9 34·2 (61)	9 30·0 (61)
66	10 05·5 (62)	9 59·8 (62)	9 54·5 (61)	9 49·6 (61)	9 44·9 (60)	9 40·3 (61)	9 36·1 (60)
67	10 11·7 (61)	10 06·0 (61)	10 00·7 (61)	9 55·7 (61)	9 50·9 (61)	9 46·4 (60)	9 42·1 (60)
68	10 17·8 (61)	10 12·1 (60)	10 06·8 (60)	10 01·7 (60)	9 57·0 (59)	9 52·4 (59)	9 48·0 (59)
69	10 23·9 (60)	10 18·1 (60)	10 12·8 (59)	10 07·7 (59)	10 02·9 (59)	9 58·3 (59)	9 53·9 (58)
70	10 29·9	10 24·1	10 18·7	10 13·6	10 08·8	10 04·2	9 59·7

TIMES OF P—continued

Depth $h =$

Δ (°)	Surface	0·00	0·01	0·02	0·03	0·04	0·05	0·06	0·07	0·08	0·09	0·10	0·11	0·12
	m s	m s	m s	m s	m s	m s	m s	m s	m s	m s	m s	m s	m s	m s
70	11 15·4	11 10·2	11 03·1	10 56·1	10 49·1	10 42·3	10 35·9	10 29·9	10 24·1	10 18·7	10 13·6	10 08·8	10 04·2	9 59·7
71	11 21·5	11 16·3	11 09·1	11 02·1	10 55·1	10 48·3	10 41·9	10 35·8	10 30·0	10 24·6	10 19·4	10 14·6	10 10·0	10 05·4
72	11 27·5	11 22·2	11 15·1	11 08·0	11 01·1	10 54·3	10 47·8	10 41·7	10 35·8	10 30·4	10 25·2	10 20·3	10 15·7	10 11·1
73	11 33·4	11 28·2	11 20·9	11 13·9	11 07·0	11 00·2	10 53·7	10 47·5	10 41·6	10 36·1	10 30·9	10 26·0	10 21·3	10 16·6
74	11 39·2	11 34·0	11 26·8	11 19·7	11 12·8	11 06·0	10 59·5	10 53·2	10 47·3	10 41·7	10 36·5	10 31·6	10 26·8	10 22·2
75	11 45·0	11 39·8	11 32·5	11 25·4	11 18·5	11 11·7	11 05·2	10 58·8	10 52·9	10 47·3	10 42·1	10 37·1	10 32·3	10 27·6
76	11 50·7	11 45·5	11 38·2	11 31·0	11 24·1	11 17·3	11 10·8	11 04·4	10 58·4	10 52·8	10 47·6	10 42·6	10 37·7	10 33·0
77	11 56·3	11 51·2	11 43·8	11 36·6	11 29·6	11 22·8	11 16·3	11 09·9	11 03·9	10 58·2	10 53·0	10 47·9	10 43·1	10 38·3
78	12 01·8	11 56·7	11 49·3	11 42·1	11 35·1	11 28·3	11 21·7	11 15·3	11 09·3	11 03·6	10 58·3	10 53·3	10 48·3	10 43·5
79	12 07·3	12 02·2	11 54·8	11 47·5	11 40·5	11 33·6	11 27·0	11 20·6	11 14·6	11 08·9	11 03·6	10 58·5	10 53·6	10 48·7
80	12 12·7	12 07·6	12 00·2	11 52·9	11 45·8	11 38·9	11 32·3	11 25·9	11 19·8	11 14·1	11 08·8	11 03·7	10 58·7	10 53·8
81	12 18·0	12 12·9	12 05·5	11 58·2	11 51·1	11 44·1	11 37·5	11 31·1	11 25·0	11 19·3	11 13·9	11 08·8	11 03·8	10 58·8
82	12 23·2	12 18·1	12 10·7	12 03·4	11 56·2	11 49·3	11 42·7	11 36·2	11 30·1	11 24·3	11 19·0	11 13·9	11 08·8	11 03·8
83	12 28·4	12 23·2	12 15·8	12 08·5	12 01·4	11 54·4	11 47·7	11 41·3	11 35·1	11 29·4	11 24·0	11 18·8	11 13·7	11 08·7
84	12 33·5	12 28·3	12 20·8	12 13·5	12 06·4	11 59·4	11 52·8	11 46·3	11 40·1	11 34·3	11 28·9	11 23·8	11 18·6	11 13·6
85	12 38·5	12 33·3	12 25·8	12 18·5	12 11·4	12 04·4	11 57·7	11 51·2	11 45·0	11 39·2	11 33·8	11 28·6	11 23·4	11 18·4
86	12 43·5	12 38·2	12 30·7	12 23·4	12 16·3	12 09·3	12 02·5	11 56·0	11 49·8	11 44·0	11 38·6	11 33·3	11 28·1	11 23·1
87	12 48·4	12 43·1	12 35·6	12 28·2	12 21·1	12 14·1	12 07·3	12 00·8	11 54·6	11 48·7	11 43·3	11 38·0	11 32·8	11 27·8
88	12 53·2	12 47·9	12 40·4	12 33·0	12 25·8	12 18·8	12 12·1	12 05·5	11 59·3	11 53·4	11 48·0	11 42·7	11 37·5	11 32·4
89	12 58·0	12 52·7	12 45·1	12 37·7	12 30·5	12 23·5	12 16·7	12 10·2	12 04·0	11 58·1	11 52·7	11 47·3	11 42·1	11 37·0
90	13 02·7	12 57·4	12 49·8	12 42·4	12 35·2	12 28·2	12 21·4	12 14·8	12 08·6	12 02·7	11 57·3	11 51·9	11 46·7	11 41·6
91	13 07·3	13 02·1	12 54·5	12 47·1	12 39·8	12 32·8	12 26·0	12 19·4	12 13·2	12 07·3	12 01·9	11 56·5	11 51·3	11 46·2
92	13 11·9	13 06·7	12 59·1	12 51·7	12 44·5	12 37·5	12 30·7	12 24·1	12 17·8	12 11·9	12 06·4	12 01·1	11 55·9	11 50·8
93	13 16·5	13 11·3	13 03·7	12 56·3	12 49·1	12 42·1	12 35·3	12 28·7	12 22·4	12 16·5	12 11·0	12 05·7	12 00·5	11 55·3
94	13 21·1	13 15·8	13 08·2	13 00·8	12 53·6	12 46·6	12 39·8	12 33·2	12 27·0	12 21·1	12 15·5	12 10·3	12 05·1	11 59·9
95	13 25·7	13 20·4	13 12·8	13 05·4	12 58·2	12 51·2	12 44·4	12 37·8	12 31·5	12 25·6	12 20·0	12 14·8	12 09·6	12 04·4
96	13 30·3	13 24·9	13 17·3	13 09·9	13 02·7	12 55·7	12 49·0	12 42·4	12 36·0	12 30·1	12 24·5	12 19·3	12 14·1	12 08·9
97	13 34·8	13 29·5	13 21·9	13 14·5	13 07·3	13 00·3	12 53·5	12 46·9	12 40·6	12 34·7	12 29·0	12 23·8	12 18·6	12 13·5
98	13 39·3	13 34·0	13 26·4	13 19·0	13 11·8	13 04·8	12 58·0	12 51·5	12 45·1	12 39·2	12 33·5	12 28·3	12 23·1	12 18·0
99	13 43·8	13 38·5	13 31·0	13 23·6	13 16·4	13 09·4	13 02·6	12 56·0	12 49·6	12 43·7	12 38·0	12 32·8	12 27·6	12 22·5
100	13 48·4	13 43·1	13 35·5	13 28·1	13 20·9	13 13·9	13 07·1	13 00·4	12 54·1	12 48·2	12 42·6	12 37·3	12 32·1	12 27·0
101	13 52·9	13 47·6	13 40·0	13 32·6	13 25·4	13 18·4	13 11·5	13 04·8	12 58·5	12 52·6	12 47·0	12 41·7	12 36·5	12 31·4
102	13 57·4	13 52·1	13 44·4	13 37·0	13 29·8	13 22·8	13 15·9	13 09·2	13 02·9	12 57·0	12 51·4	12 46·1	12 40·9	12 35·8
103	14 01·8	13 56·5	13 48·8	13 41·4	13 34·2	13 27·2	13 20·3	13 13·6	13 07·3	13 01·4				
104	14 06·2	14 00·9	13 53·2	13 45·8	13 38·6	13 31·6								
105	14 10·6	14 05·3												

TIMES OF S

Each cell gives the time as **m s** (minutes, seconds); the small number following each value is the first difference to the line below.

Depth $h =$ (Surface and 0·00–0·05)

Δ	Surface	0·00	0·01	0·02	0·03	0·04	0·05
0·0	(10·7) 127	9·2 89	23·6 39	37·7 23	51·5 15	1 05·1 12	1 18·3 9
0·5	(23·4) 127	(18·1) 127	27·5 83	40·0 57	53·0 44	1 06·3 34	1 19·2 27
1·0	36·1 127	30·8 127	35·8 108	45·7 84	57·4 65	1 09·7 51	1 21·9 44
1·5	48·8 127	43·5 126	46·6 106	54·1 97	1 03·9 80	1 14·8 67	1 26·3 57
2·0	1 01·5 127	56·1 127	57·2 123	1 03·8 106	1 11·9 92	1 21·5 80	1 32·0 67
2·5	1 14·2 127	1 08·8 127	1 09·5 123	1 14·4 111	1 21·1 100	1 29·5 87	1 38·7 78
3·0	1 26·9 127	1 21·5 127	1 21·8 123	1 25·5 114	1 31·1 103	1 38·2 94	1 46·5 84
3·5	1 39·6 126	1 34·2 127	1 34·1 123	1 36·9 116	1 41·4 108	1 47·6 99	1 54·9 90
4·0	1 52·2 127	1 46·9 126	1 46·4 123	1 48·5 116	1 52·2 110	1 57·5 103	2 03·9 94
4·5	2 04·9 126	1 59·5 126	1 58·7 124	2 00·1 118	2 03·2 112	2 07·8 104	2 13·3 97
5·0	2 17·5 126	2 12·1 126	2 11·1 123	2 11·9 119	2 14·4 113	2 18·2 107	2 23·0 100
5·5	2 30·1 125	2 24·7 125	2 23·4 123	2 23·8 119	2 25·7 114	2 28·9 108	2 33·0 102
6·0	2 42·6 125	2 37·2 125	2 35·7 123	2 35·7 119	2 37·1 114	2 39·7 108	2 43·2 102
6·5	2 55·1 125	2 49·7 125	2 48·0 123	2 47·6 120	2 48·6 115	2 50·7 110	2 53·6 104
7·0	3 07·6 125	3 02·1 124	3 00·2 123	2 59·6 120	3 00·1 115	3 01·7 111	3 04·1 106
7·5	3 20·1 125	3 14·6 124	3 12·5 122	3 11·6 119	3 11·6 116	3 12·8 111	3 14·7 107
8·0	3 32·6 125	3 27·0 125	3 24·7 122	3 23·5 119	3 23·2 115	3 23·9 111	3 25·4 107
8·5	3 45·1 125	3 39·5 124	3 36·9 122	3 35·4 118	3 34·7 115	3 35·0 112	3 36·1 108
9·0	3 57·5 124	3 51·9 124	3 49·1 121	3 47·2 118	3 46·2 115	3 46·2 112	3 46·9 108
9·5	4 09·9 123	4 04·3 123	4 01·2 120	3 59·0 118	3 57·7 115	3 57·4 112	3 57·7 109
10·0	4 22·2 245	4 16·6 244	4 13·2 240	4 10·8 235	4 09·2 231	4 08·6 224	4 08·6 216
11·0	4 46·7 244	4 41·0 243	4 37·2 239	4 34·3 234	4 32·3 229	4 31·0 223	4 30·2 216
12·0	5 11·1 242	5 05·3 241	5 01·1 237	4 57·7 233	4 55·2 227	4 53·3 221	4 51·8 216
13·0	5 35·3 239	5 29·4 239	5 24·8 235	5 21·0 230	5 17·9 224	5 15·4 219	5 13·4 216
14·0	5 59·2 237	5 53·3 236	5 48·3 235	5 44·0 228	5 40·3 223	5 37·3 218	d5 35·0 202
15·0	6 22·9 235	6 16·9 233	6 11·8 228	6 06·8 226	6 02·6 222	5 59·1 215	r5 55·2 193
16·0	6 46·4 231	6 40·2 231	6 34·6 227	6 29·4 223	6 24·8 217	d6 20·6 196	6 14·5 190
17·0	7 09·5 228	7 03·3 227	6 57·3 221	6 51·7 219	d6 46·5 202	r6 40·2 191	6 33·5 188
18·0	7 32·3 226	7 26·0 225	7 19·4 220	d7 13·6 204	r7 06·7 191	6 59·3 187	6 52·3 183
19·0	7 54·9 222	7 48·5 221	d7 41·4 202	r7 34·0 187	7 25·8 185	7 18·0 181	7 10·6 177
20·0	d8 17·1 203	d8 10·6 194	r8 01·6 189	7 52·7 186	7 44·3 180	7 36·1 176	7 28·3 172
21·0	r8 37·4 188	r8 30·0 187	8 20·5 182	8 11·3 178	8 02·3 175	7 53·7 172	7 45·5 170
22·0	8 56·2 182	8 48·7 181	8 38·7 176	8 29·1 172	8 19·8 170	8 10·9 169	8 02·5 166
23·0	9 14·4 175	9 06·8 175	8 56·3 172	8 46·3 169	8 36·8 168	8 27·8 166	8 19·1 165
24·0	9 31·9 170	9 24·3 170	9 13·5 168	9 03·2 168	8 53·6 165	8 44·4 163	8 35·6 162
25·0	9 48·9 167	9 41·3 165	9 30·3 166	9 20·0 165	9 10·1 162	9 00·7 161	8 51·8 160
26·0	10 05·6 165	9 57·8 164	9 46·9 163	9 36·5 163	9 26·3 162	9 16·8 161	9 07·8 160
27·0	10 22·1 161	10 14·2 161	10 03·2 162	9 52·7 162	9 42·5 160	9 32·8 160	9 23·7 159
28·0	10 38·2 161	10 30·3 160	10 19·4 159	10 08·8 159	9 58·5 159	9 48·7 159	9 39·5 158
29·0	10 54·3 159	10 46·3 159	10 35·3 158	10 24·7 158	10 14·4 157	10 04·5 157	9 55·2 157
30·0	11 10·2	11 02·2	10 51·1	10 40·4	10 30·1	10 20·2	10 10·9

TIMES OF S

Depth $h =$ (0·06–0·12)

Δ	0·06	0·07	0·08	0·09	0·10	0·11	0·12
0·0	1 31·2 8	1 43·6 7	1 55·5 5	2 06·9 5	2 17·9 4	2 28·6 4	2 39·0 4
0·5	1 32·0 22	1 44·3 20	1 56·0 16	2 07·4 15	2 18·3 12	2 29·0 12	2 39·4 10
1·0	1 34·2 37	1 46·3 31	1 57·6 27	2 08·9 23	2 19·6 21	2 30·2 19	2 40·4 17
1·5	1 37·9 49	1 49·4 42	2 00·3 38	2 11·2 32	2 21·7 29	2 32·1 25	2 42·1 23
2·0	1 42·8 59	1 53·6 52	2 04·1 46	2 14·4 40	2 24·6 36	2 34·6 32	2 44·4 29
2·5	1 48·7 68	1 58·8 60	2 08·7 53	2 18·4 48	2 28·2 42	2 37·8 38	2 47·3 34
3·0	1 55·5 75	2 04·8 67	2 14·0 60	2 23·2 54	2 32·4 48	2 41·6 44	2 50·7 39
3·5	2 03·0 81	2 11·5 73	2 20·0 66	2 28·6 60	2 37·2 54	2 46·0 49	2 54·6 45
4·0	2 11·1 86	2 18·8 78	2 26·6 70	2 34·6 64	2 42·6 58	2 50·9 53	2 59·1 50
4·5	2 19·7 89	2 26·6 83	2 33·6 76	2 41·0 68	2 48·4 64	2 56·2 58	3 04·1 53
5·0	2 28·6 92	2 34·9 86	2 41·2 79	2 47·8 73	2 54·8 65	3 02·0 61	3 09·4 56
5·5	2 37·8 96	2 43·5 89	2 49·1 83	2 55·1 76	3 01·3 70	3 08·1 65	3 15·0 59
6·0	2 47·4 100	2 52·4 92	2 57·4 85	3 02·7 79	3 08·3 73	3 14·6 63	3 20·9 61
6·5	2 57·4 101	3 01·6 92	3 05·9 87	3 10·6 79	3 15·6 75	3 20·9 73	3 27·0 64
7·0	3 07·5 101	3 10·9 95	3 14·6 89	3 18·7 84	3 23·1 77	3 28·2 72	3 33·4 66
7·5	3 17·6 101	3 20·4 96	3 23·5 90	3 27·1 85	3 30·8 80	3 35·4 74	3 40·0 69
8·0	3 27·7 103	3 30·0 96	3 32·5 91	3 35·6 86	3 38·8 81	3 42·8 76	3 46·9 71
8·5	3 38·0 103	3 39·7 97	3 41·6 92	3 44·2 87	3 46·9 83	3 50·4 77	3 54·0 74
9·0	3 48·3 104	3 49·4 97	3 50·8 92	3 52·9 88	3 55·2 83	3 58·1 78	4 01·4 74
9·5	3 58·7 104	3 59·2 98	4 00·1 93	4 01·7 88	4 03·5 83	4 05·9 79	4 08·9 76
10·0	4 09·1 209	4 09·0 197	4 09·4 187	4 10·5 178	4 11·9 170	4 13·8 162	4 16·5 154
11·0	d4 30·0 209	4 28·7 196	4 28·1 187	4 28·3 177	4 28·9 170	4 30·0 164	4 31·9 156
12·0	r4 50·9 201	4 48·3 193	4 46·8 184	4 46·0 177	4 45·9 171	4 46·4 164	4 47·5 158
13·0	5 11·0 198	5 07·6 191	5 05·2 184	5 03·7 178	5 03·0 171	5 02·8 165	5 03·3 159
14·0	5 30·8 193	5 26·7 189	5 23·6 183	5 21·5 176	5 20·1 169	5 19·3 164	5 19·2 159
15·0	5 50·1 189	5 45·6 184	5 41·9 179	5 39·1 173	5 37·0 167	5 35·7 162	5 35·1 159
16·0	6 09·0 186	6 04·0 181	5 59·8 176	5 56·4 171	5 53·7 167	5 51·9 163	5 51·0 159
17·0	6 27·6 182	6 22·1 176	6 17·4 171	6 13·5 168	6 10·4 165	6 08·2 162	6 06·9 158
18·0	6 45·8 177	6 39·7 173	6 34·5 170	6 30·3 167	6 26·9 163	6 24·4 160	6 22·7 158
19·0	7 03·5 173	6 57·0 170	6 51·5 167	6 47·0 165	6 43·2 163	6 40·4 159	6 38·5 156
20·0	7 20·8 170	7 14·0 168	7 08·2 166	7 03·5 162	6 59·4 162	6 56·3 159	6 54·1 156
21·0	7 37·8 167	7 30·8 165	7 24·8 163	7 19·7 162	7 15·6 159	7 12·2 158	7 09·7 156
22·0	7 54·5 165	7 47·3 163	7 41·1 161	7 35·9 159	7 31·5 158	7 28·0 157	7 25·3 155
23·0	8 11·0 162	8 03·6 161	7 57·2 160	7 51·8 158	7 47·2 157	7 43·6 156	7 40·8 155
24·0	8 27·2 161	8 19·7 160	8 13·2 158	8 07·6 157	8 02·9 156	7 59·1 155	7 56·3 155
25·0	8 43·3 160	8 35·7 159	8 29·0 158	8 23·3 157	8 18·5 156	8 14·7 155	8 11·8 154
26·0	8 59·3 158	8 51·6 157	8 44·8 156	8 39·0 155	8 34·1 155	8 30·2 155	8 27·2 153
27·0	9 15·1 157	9 07·3 156	9 00·4 156	8 54·5 156	8 49·6 155	8 45·6 154	8 42·5 153
28·0	9 30·8 156	9 22·9 156	9 16·0 155	9 10·1 154	9 05·1 154	9 01·0 153	8 57·8 151
29·0	9 46·4 156	9 38·5 155	9 31·5 155	9 25·5 155	9 20·5 154	9 16·3 153	9 12·9 151
30·0	10 02·0	9 54·0	9 47·0	9 41·0	9 35·9	9 31·6	9 28·0

TIMES OF S—*continued*

Depth $h =$

Values are given as minutes and seconds (m s); the small number in parentheses following each value is the first difference to the next line.

Δ (°)	Surface	0·00	0·01	0·02	0·03	0·04	0·05
30	11 10·2 (158)	11 02·2 (157)	10 51·1 (157)	10 40·4 (157)	10 30·1 (157)	10 20·2 (156)	10 10·9 (156)
31	11 26·0 (156)	11 17·9 (157)	11 06·8 (157)	10 56·1 (156)	10 45·8 (155)	10 35·8 (155)	10 26·5 (155)
32	11 41·6 (156)	11 33·6 (156)	11 22·5 (155)	11 11·7 (156)	11 01·3 (155)	10 51·3 (155)	10 42·0 (154)
33	11 57·2 (155)	11 49·2 (155)	11 38·0 (155)	11 27·3 (154)	11 16·8 (155)	11 06·8 (154)	10 57·4 (153)
34	12 12·7 (155)	12 04·7 (155)	11 53·5 (154)	11 42·7 (154)	11 32·3 (153)	11 22·2 (153)	11 12·7 (153)
35	12 28·2 (155)	12 20·2 (154)	12 08·9 (154)	11 58·1 (153)	11 47·6 (153)	11 37·5 (152)	11 28·0 (152)
36	12 43·7 (153)	12 35·6 (154)	12 24·3 (152)	12 13·4 (152)	12 02·9 (152)	11 52·7 (152)	11 43·2 (150)
37	12 59·0 (153)	12 51·0 (152)	12 39·5 (152)	12 28·6 (152)	12 18·1 (150)	12 07·9 (150)	11 58·2 (150)
38	13 14·3 (151)	13 06·2 (152)	12 54·7 (151)	12 43·8 (150)	12 33·1 (150)	12 22·9 (150)	12 13·2 (148)
39	13 29·4 (151)	13 21·4 (150)	13 09·8 (150)	12 58·8 (149)	12 48·1 (149)	12 37·9 (148)	12 28·0 (148)
40	13 44·5 (149)	13 36·4 (149)	13 24·8 (148)	13 13·7 (148)	13 03·0 (147)	12 52·7 (147)	12 42·8 (146)
41	13 59·4 (148)	13 51·3 (148)	13 39·6 (148)	13 28·5 (146)	13 17·7 (146)	13 07·4 (145)	12 57·4 (145)
42	14 14·2 (147)	14 06·1 (146)	13 54·4 (146)	13 43·1 (146)	13 32·3 (145)	13 21·9 (145)	13 11·9 (144)
43	14 28·9 (145)	14 20·7 (146)	14 09·0 (145)	13 57·7 (145)	13 46·8 (144)	13 36·4 (144)	13 26·3 (143)
44	14 43·4 (145)	14 35·3 (144)	14 23·5 (144)	14 12·2 (143)	14 01·2 (143)	13 50·8 (142)	13 40·6 (142)
45	14 57·9 (143)	14 49·7 (143)	14 37·9 (143)	14 26·5 (142)	14 15·5 (142)	14 05·0 (142)	13 54·8 (141)
46	15 12·2 (143)	15 04·0 (142)	14 52·2 (142)	14 40·7 (142)	14 29·7 (142)	14 19·2 (141)	14 08·9 (140)
47	15 26·5 (141)	15 18·2 (142)	15 06·4 (140)	14 54·9 (140)	14 43·8 (141)	14 33·2 (140)	14 22·9 (140)
48	15 40·6 (141)	15 32·4 (140)	15 20·4 (140)	15 08·9 (140)	14 57·9 (139)	14 47·2 (140)	14 36·9 (138)
49	15 54·7 (139)	15 46·4 (139)	15 34·4 (139)	15 22·9 (138)	15 11·8 (138)	15 01·0 (138)	14 50·7 (137)
50	16 08·6 (138)	16 00·3 (138)	15 48·3 (138)	15 36·7 (138)	15 25·6 (137)	15 14·8 (137)	15 04·4 (136)
51	16 22·4 (138)	16 14·1 (138)	16 02·1 (137)	15 50·5 (136)	15 39·3 (136)	15 28·5 (135)	15 18·0 (135)
52	16 36·2 (136)	16 27·9 (136)	16 15·8 (136)	16 04·1 (136)	15 52·9 (135)	15 42·0 (135)	15 31·5 (135)
53	16 49·8 (136)	16 41·5 (136)	16 29·4 (135)	16 17·7 (134)	16 06·4 (134)	15 55·5 (133)	15 45·0 (133)
54	17 03·4 (134)	16 55·1 (134)	16 42·9 (134)	16 31·1 (134)	16 19·8 (133)	16 08·8 (133)	15 58·3 (132)
55	17 16·8 (134)	17 08·5 (134)	16 56·3 (133)	16 44·5 (133)	16 33·1 (132)	16 22·1 (132)	16 11·5 (131)
56	17 30·2 (132)	17 21·9 (132)	17 09·6 (132)	16 57·8 (131)	16 46·3 (132)	16 35·3 (131)	16 24·6 (131)
57	17 43·4 (132)	17 35·1 (131)	17 22·8 (131)	17 10·9 (131)	16 59·5 (130)	16 48·4 (130)	16 37·7 (129)
58	17 56·6 (131)	17 48·2 (131)	17 35·9 (130)	17 24·0 (130)	17 12·5 (130)	17 01·4 (129)	16 50·6 (129)
59	18 09·7 (129)	18 01·3 (129)	17 48·9 (129)	17 37·0 (128)	17 25·5 (128)	17 14·3 (128)	17 03·5 (127)
60	18 22·6 (128)	18 14·2 (128)	18 01·8 (128)	17 49·8 (127)	17 38·3 (127)	17 27·1 (126)	17 16·2 (126)
61	18 35·4 (127)	18 27·0 (127)	18 14·6 (126)	18 02·5 (126)	17 51·0 (126)	17 39·7 (125)	17 28·8 (125)
62	18 48·1 (126)	18 39·7 (125)	18 27·2 (126)	18 15·1 (125)	18 03·5 (125)	17 52·2 (124)	17 41·3 (123)
63	19 00·7 (125)	18 52·2 (125)	18 39·8 (124)	18 27·6 (124)	18 16·0 (123)	18 04·6 (123)	17 53·6 (123)
64	19 13·2 (123)	19 04·7 (123)	18 52·2 (123)	18 40·0 (123)	18 28·3 (122)	18 16·9 (122)	18 05·9 (121)
65	19 25·5 (123)	19 17·0 (122)	19 04·5 (122)	18 52·3 (122)	18 40·5 (121)	18 29·1 (121)	18 18·0 (121)
66	19 37·8 (121)	19 29·2 (121)	19 16·7 (121)	19 04·5 (120)	18 52·6 (121)	18 41·2 (120)	18 30·1 (119)
67	19 49·9 (120)	19 41·3 (120)	19 28·8 (120)	19 16·5 (119)	19 04·7 (119)	18 53·2 (118)	18 42·0 (118)
68	20 01·9 (119)	19 53·3 (119)	19 40·8 (118)	19 28·4 (119)	19 16·6 (117)	19 05·0 (118)	18 53·8 (117)
69	20 13·8 (118)	20 05·2 (118)	19 52·7 (117)	19 40·2 (117)	19 28·3 (117)	19 16·8 (116)	19 05·5 (116)
70	20 25·6	20 17·0	20 04·4	19 51·9	19 40·0	19 28·4	19 17·1

Δ (°)	0·06	0·07	0·08	0·09	0·10	0·11	0·12
30	10 02·0 (155)	9 54·0 (155)	9 47·0 (155)	9 41·0 (154)	9 35·9 (152)	9 31·6 (151)	9 28·0 (150)
31	10 17·5 (155)	10 09·5 (154)	10 02·5 (153)	9 56·4 (152)	9 51·1 (152)	9 46·7 (150)	9 43·0 (149)
32	10 33·0 (154)	10 24·9 (154)	10 17·8 (154)	10 11·6 (152)	10 06·3 (151)	10 01·7 (150)	9 57·9 (148)
33	10 48·4 (153)	10 40·3 (153)	10 33·2 (152)	10 26·8 (151)	10 21·4 (149)	10 16·7 (148)	10 12·7 (147)
34	11 03·7 (152)	10 55·6 (152)	10 48·4 (151)	10 41·9 (150)	10 36·3 (149)	10 31·5 (147)	10 27·4 (146)
35	11 18·9 (151)	11 10·8 (150)	11 03·5 (149)	10 56·9 (149)	10 51·2 (147)	10 46·2 (146)	10 42·0 (145)
36	11 34·0 (150)	11 25·8 (150)	11 18·4 (149)	11 11·8 (147)	11 05·9 (147)	11 00·8 (146)	10 56·5 (144)
37	11 49·0 (149)	11 40·8 (148)	11 33·3 (147)	11 26·5 (146)	11 20·6 (145)	11 15·4 (144)	11 10·9 (143)
38	12 03·9 (148)	11 55·6 (147)	11 48·0 (146)	11 41·1 (146)	11 35·1 (145)	11 29·8 (144)	11 25·2 (142)
39	12 18·7 (147)	12 10·3 (146)	12 02·6 (145)	11 55·7 (144)	11 49·6 (143)	11 44·2 (142)	11 39·4 (141)
40	12 33·4 (146)	12 24·9 (145)	12 17·1 (144)	12 10·1 (143)	12 03·9 (142)	11 58·4 (141)	11 53·5 (141)
41	12 48·0 (144)	12 39·4 (143)	12 31·5 (143)	12 24·4 (142)	12 18·1 (142)	12 12·5 (141)	12 07·6 (139)
42	13 02·4 (143)	12 53·7 (143)	12 45·8 (142)	12 38·6 (141)	12 32·3 (140)	12 26·6 (140)	12 21·5 (139)
43	13 16·7 (143)	13 08·0 (141)	13 00·0 (141)	12 52·7 (141)	12 46·3 (139)	12 40·6 (138)	12 35·4 (138)
44	13 31·0 (141)	13 22·1 (141)	13 14·1 (140)	13 06·8 (139)	13 00·2 (139)	12 54·4 (138)	12 49·2 (137)
45	13 45·1 (140)	13 36·2 (140)	13 28·1 (139)	13 20·7 (139)	13 14·1 (138)	13 08·2 (137)	13 02·9 (136)
46	13 59·1 (140)	13 50·2 (139)	13 42·0 (139)	13 34·6 (138)	13 27·9 (137)	13 21·9 (136)	13 16·5 (135)
47	14 13·1 (138)	14 04·1 (138)	13 55·9 (137)	13 48·4 (137)	13 41·6 (136)	13 35·5 (135)	13 30·0 (134)
48	14 26·9 (137)	14 17·9 (137)	14 09·6 (136)	14 02·1 (136)	13 55·2 (136)	13 49·0 (135)	13 43·4 (134)
49	14 40·6 (137)	14 31·6 (136)	14 23·2 (136)	14 15·7 (135)	14 08·8 (134)	14 02·5 (133)	13 56·8 (132)
50	14 54·3 (136)	14 45·2 (135)	14 36·8 (135)	14 29·2 (134)	14 22·2 (133)	14 15·8 (132)	14 10·0 (132)
51	15 07·9 (135)	14 58·7 (135)	14 50·3 (134)	14 42·6 (133)	14 35·5 (132)	14 29·0 (132)	14 23·2 (130)
52	15 21·4 (134)	15 12·2 (133)	15 03·7 (132)	14 55·9 (132)	14 48·7 (132)	14 42·2 (130)	14 36·2 (130)
53	15 34·8 (133)	15 25·5 (133)	15 16·9 (132)	15 09·1 (131)	15 01·9 (130)	14 55·2 (130)	14 49·2 (128)
54	15 48·1 (132)	15 38·8 (131)	15 30·1 (131)	15 22·2 (130)	15 14·9 (129)	15 08·2 (128)	15 02·0 (128)
55	16 01·3 (131)	15 51·9 (130)	15 43·2 (130)	15 35·2 (129)	15 27·8 (128)	15 21·0 (127)	15 14·8 (126)
56	16 14·4 (130)	16 04·9 (130)	15 56·2 (128)	15 48·1 (128)	15 40·6 (128)	15 33·7 (127)	15 27·4 (126)
57	16 27·4 (129)	16 17·9 (128)	16 09·0 (128)	16 00·9 (127)	15 53·4 (126)	15 46·4 (125)	15 40·0 (124)
58	16 40·3 (127)	16 30·7 (127)	16 21·8 (127)	16 13·6 (126)	16 06·0 (126)	15 58·9 (125)	15 52·4 (123)
59	16 53·0 (127)	16 43·4 (126)	16 34·5 (125)	16 26·2 (125)	16 18·6 (124)	16 11·4 (123)	16 04·7 (122)
60	17 05·7 (125)	16 56·0 (125)	16 47·0 (124)	16 38·7 (124)	16 31·0 (123)	16 23·7 (122)	16 16·9 (121)
61	17 18·2 (125)	17 08·5 (124)	16 59·4 (123)	16 51·1 (122)	16 43·3 (122)	16 35·9 (121)	16 29·0 (121)
62	17 30·7 (123)	17 20·9 (122)	17 11·7 (122)	17 03·3 (122)	16 55·5 (120)	16 48·0 (120)	16 41·1 (119)
63	17 43·0 (122)	17 33·1 (122)	17 23·9 (121)	17 15·5 (120)	17 07·5 (120)	17 00·0 (119)	16 53·0 (118)
64	17 55·2 (121)	17 45·3 (120)	17 36·0 (120)	17 27·5 (119)	17 19·5 (118)	17 11·9 (118)	17 04·8 (117)
65	18 07·3 (120)	17 57·3 (119)	17 48·0 (119)	17 39·4 (118)	17 31·3 (117)	17 23·7 (116)	17 16·5 (116)
66	18 19·3 (119)	18 09·2 (119)	17 59·9 (117)	17 51·2 (117)	17 43·0 (116)	17 35·3 (116)	17 28·1 (114)
67	18 31·2 (117)	18 21·1 (117)	18 11·6 (117)	18 02·9 (115)	17 54·6 (115)	17 46·9 (114)	17 39·5 (114)
68	18 42·9 (117)	18 32·8 (115)	18 23·3 (115)	18 14·4 (115)	18 06·1 (114)	17 58·3 (113)	17 50·9 (112)
69	18 54·6 (115)	18 44·3 (115)	18 34·8 (114)	18 25·9 (113)	18 17·5 (113)	18 09·6 (112)	18 02·1 (111)
70	19 06·1	18 55·8	18 46·2	18 37·2	18 28·8	18 20·8	18 13·2

TIMES OF S—*continued*

Depth $h =$

Values are minutes and seconds (m s); the number in parentheses after a value is the interpolation difference printed below it.

Δ°	Surface	0·00	0·01	0·02	0·03	0·04	0·05
70	20 25·6 (117)	20 17·0 (116)	20 04·4 (115)	19 51·9 (115)	19 40·0 (115)	19 28·4 (115)	19 17·1 (114)
71	20 37·3 (115)	20 28·6 (116)	20 15·9 (115)	20 03·4 (114)	19 51·5 (114)	19 39·9 (113)	19 28·5 (113)
72	20 48·8 (114)	20 40·2 (114)	20 27·4 (113)	20 14·8 (113)	20 02·9 (112)	19 51·2 (112)	19 39·8 (112)
73	21 00·2 (112)	20 51·6 (112)	20 38·7 (111)	20 26·1 (112)	20 14·1 (111)	20 02·4 (110)	19 51·0 (110)
74	21 11·4 (112)	21 02·8 (112)	20 49·8 (111)	20 37·3 (110)	20 25·2 (110)	20 13·4 (110)	20 02·0 (109)
75	21 22·6 (110)	21 14·0 (110)	21 00·9 (110)	20 48·3 (110)	20 36·2 (109)	20 24·4 (109)	20 12·9 (108)
76	21 33·6 (109)	21 25·0 (109)	21 11·9 (109)	20 59·3 (108)	20 47·1 (108)	20 35·3 (107)	20 23·7 (107)
77	21 44·5 (108)	21 35·9 (108)	21 22·8 (107)	21 10·1 (107)	20 57·9 (106)	20 46·0 (106)	20 34·4 (106)
78	21 55·3 (107)	21 46·7 (106)	21 33·5 (107)	21 20·8 (106)	21 08·5 (106)	20 56·6 (105)	20 45·0 (105)
79	22 06·0 (105)	21 57·3 (105)	21 44·2 (105)	21 31·4 (105)	21 19·1 (104)	21 07·1 (104)	20 55·5 (103)
80	22 16·5 (104)	22 07·8 (104)	21 54·7 (104)	21 41·9 (103)	21 29·5 (103)	21 17·5 (102)	21 05·8 (102)
81	22 26·9 (103)	22 18·2 (103)	22 05·1 (102)	21 52·2 (102)	21 39·8 (101)	21 27·7 (101)	21 16·0 (100)
82	22 37·2 (102)	22 28·5 (101)	22 15·3 (101)	22 02·4 (101)	21 49·9 (101)	21 37·8 (100)	21 26·0 (100)
83	22 47·4 (100)	22 38·6 (100)	22 25·4 (100)	22 12·5 (99)	22 00·0 (99)	21 47·8 (98)	21 36·0 (97)
84	22 57·4 (99)	22 48·6 (99)	22 35·4 (98)	22 22·4 (98)	22 09·9 (97)	21 57·6 (97)	21 45·7 (97)
85	23 07·3 (97)	22 58·5 (97)	22 45·2 (97)	22 32·2 (96)	22 19·6 (96)	22 07·3 (95)	21 55·4 (95)
86	23 17·0 (96)	23 08·2 (96)	22 54·9 (95)	22 41·8 (95)	22 29·2 (94)	22 16·8 (95)	22 04·9 (94)
87	23 26·6 (94)	23 17·8 (94)	23 04·4 (94)	22 51·3 (94)	22 38·6 (94)	22 26·3 (92)	22 14·3 (92)
88	23 36·0 (93)	23 27·2 (93)	23 13·8 (93)	23 00·7 (92)	22 48·0 (91)	22 35·5 (92)	22 23·5 (91)
89	23 45·3 (92)	23 36·5 (92)	23 23·1 (91)	23 09·9 (91)	22 57·1 (91)	22 44·7 (90)	22 32·6 (90)
90	23 54·5 (89)	23 45·7 (89)	23 32·2 (89)	23 19·0 (89)	23 06·2 (89)	22 53·7 (88)	22 41·6 (88)
91	24 03·4 (89)	23 54·6 (88)	23 41·1 (88)	23 27·9 (88)	23 15·1 (87)	23 02·5 (88)	22 50·4 (87)
92	24 12·3 (87)	24 03·4 (88)	23 49·9 (87)	23 36·7 (87)	23 23·8 (87)	23 11·3 (86)	22 59·1 (87)
93	24 21·0 (87)	24 12·2 (86)	23 58·6 (86)	23 45·4 (86)	23 32·5 (86)	23 19·9 (86)	23 07·8 (85)
94	24 29·7 (85)	24 20·8 (85)	24 07·2 (85)	23 54·0 (85)	23 41·1 (85)	23 28·5 (85)	23 16·3 (85)
95	24 38·2 (85)	24 29·3 (85)	24 15·7 (85)	24 02·5 (85)	23 49·6 (85)	23 37·0 (85)	23 24·8 (85)
96	24 46·7 (85)	24 37·8 (85)	24 24·2 (85)	24 11·0 (85)	23 58·1 (85)	23 45·5 (84)	23 33·3 (84)
97	24 55·2 (84)	24 46·3 (85)	24 32·7 (84)	24 19·5 (84)	24 06·6 (84)	23 53·9 (85)	23 41·7 (85)
98	25 03·6 (84)	24 54·7 (84)	24 41·1 (84)	24 27·9 (83)	24 15·0 (83)	24 02·4 (83)	23 50·2 (83)
99	25 12·0 (84)	25 03·1 (84)	24 49·5 (84)	24 36·2 (84)	24 23·3 (84)	24 10·7 (84)	23 58·5 (84)
100	25 20·4 (84)	25 11·5 (84)	24 57·9 (84)	24 44·6 (84)	24 31·7 (84)	24 19·1 (83)	24 06·9 (84)
101	25 28·8 (83)	25 19·9 (83)	25 06·3 (83)	24 53·0 (83)	24 40·1 (83)	24 27·4 (84)	24 15·3 (83)
102	25 37·1 (84)	25 28·2 (84)	25 14·6 (84)	25 01·3 (84)	24 48·4 (84)	24 35·8 (83)	24 23·6 (83)
103	25 45·5 (83)	25 36·6 (83)	25 23·0 (83)	25 09·7 (83)	24 56·8 (83)	24 44·1 (83)	24 31·9 (83)
104	25 53·8 (83)	25 44·9 (83)	25 31·3 (83)	25 18·0 (83)	25 05·1 (82)	24 52·4 (83)	24 40·2 (83)
105	26 02·1 (83)	25 53·2 (83)	25 39·6 (82)	25 26·3 (83)	25 13·3 (84)	25 00·7 (83)	24 48·5
106	26 10·4 (83)	26 01·5 (83)	25 47·8 (84)	25 34·6	25 21·7	25 09·0	
107	26 18·7	26 09·8	25 56·2				

Δ°	0·06	0·07	0·08	0·09	0·10	0·11	0·12
70	19 06·1 (114)	18 55·8 (113)	18 46·2 (113)	18 37·2 (112)	18 28·8 (112)	18 20·8 (111)	18 13·2 (110)
71	19 17·5 (114)	19 07·1 (113)	18 57·5 (113)	18 48·4 (112)	18 40·0 (112)	18 31·9 (111)	18 24·2 (110)
72	19 28·7 (112)	19 18·3 (112)	19 08·6 (111)	18 59·5 (111)	18 51·1 (111)	18 42·9 (110)	18 35·1 (109)
73	19 39·8 (110)	19 29·4 (111)	19 19·6 (110)	19 10·5 (110)	19 02·0 (109)	18 53·7 (108)	18 45·9 (108)
74	19 50·8 (110)	19 40·4 (108)	19 30·5 (109)	19 21·4 (109)	19 12·9 (107)	19 04·5 (106)	18 56·5 (106)
75	20 01·7 (108)	19 51·2 (107)	19 41·3 (107)	19 32·2 (107)	19 23·6 (106)	19 15·1 (105)	19 07·1 (105)
76	20 12·5 (106)	20 01·9 (107)	19 52·0 (106)	19 42·9 (105)	19 34·2 (104)	19 25·6 (105)	19 17·6 (103)
77	20 23·1 (106)	20 12·6 (105)	20 02·6 (104)	19 53·4 (104)	19 44·6 (103)	19 36·1 (103)	19 27·9 (103)
78	20 33·7 (104)	20 23·1 (103)	20 13·0 (104)	20 03·8 (103)	19 54·9 (102)	19 46·4 (102)	19 38·2 (101)
79	20 44·1 (103)	20 33·4 (103)	20 23·4 (102)	20 14·1 (101)	20 05·1 (101)	19 56·6 (100)	19 48·3 (100)
80	20 54·4 (102)	20 43·7 (101)	20 33·6 (101)	20 24·2 (100)	20 15·2 (99)	20 06·6 (99)	19 58·3 (98)
81	21 04·6 (100)	20 53·8 (100)	20 43·7 (99)	20 34·2 (98)	20 25·1 (99)	20 16·5 (97)	20 08·1 (98)
82	21 14·6 (100)	21 03·8 (99)	20 53·6 (98)	20 44·0 (98)	20 35·0 (96)	20 26·2 (97)	20 17·9 (95)
83	21 24·6 (98)	21 13·7 (97)	21 03·4 (97)	20 53·8 (95)	20 44·6 (96)	20 35·9 (94)	20 27·4 (95)
84	21 34·4 (96)	21 23·4 (96)	21 13·1 (95)	21 03·3 (95)	20 54·2 (94)	20 45·3 (94)	20 36·9 (93)
85	21 44·0 (95)	21 33·0 (94)	21 22·6 (94)	21 12·8 (93)	21 03·6 (92)	20 54·7 (92)	20 46·2 (91)
86	21 53·5 (93)	21 42·4 (93)	21 32·0 (92)	21 22·1 (92)	21 12·8 (91)	21 03·9 (91)	20 55·3 (91)
87	22 02·8 (93)	21 51·7 (93)	21 41·2 (91)	21 31·3 (92)	21 21·9 (90)	21 13·0 (89)	21 04·4 (89)
88	22 12·0 (91)	22 00·9 (90)	21 50·3 (90)	21 40·4 (89)	21 30·9 (89)	21 21·9 (89)	21 13·3 (88)
89	22 21·1 (89)	22 09·9 (89)	21 59·3 (89)	21 49·3 (89)	21 39·8 (88)	21 30·8 (87)	21 22·1 (87)
90	22 30·0 (88)	22 18·8 (88)	22 08·2 (87)	21 58·2 (87)	21 48·6 (87)	21 39·5 (86)	21 30·8 (86)
91	22 38·8 (87)	22 27·6 (86)	22 16·9 (87)	22 06·9 (86)	21 57·3 (86)	21 48·1 (86)	21 39·4 (85)
92	22 47·5 (86)	22 36·2 (86)	22 25·6 (86)	22 15·5 (85)	22 05·9 (85)	21 56·7 (85)	21 47·9 (85)
93	22 56·1 (85)	22 44·8 (85)	22 34·2 (85)	22 24·1 (85)	22 14·4 (85)	22 05·2 (84)	21 56·4 (84)
94	23 04·6 (85)	22 53·3 (85)	22 42·7 (84)	22 32·6 (84)	22 22·9 (84)	22 13·6 (84)	22 04·8 (84)
95	23 13·1 (85)	23 01·8 (85)	22 51·1 (84)	22 41·0 (84)	22 31·3 (84)	22 22·0 (84)	22 13·2 (84)
96	23 21·6 (84)	23 10·3 (84)	22 59·5 (84)	22 49·4 (84)	22 39·7 (84)	22 30·4 (83)	22 21·6 (84)
97	23 30·0 (84)	23 18·7 (84)	23 07·9 (84)	22 57·8 (83)	22 48·1 (83)	22 38·7 (84)	22 29·9 (84)
98	23 38·4 (84)	23 27·1 (84)	23 16·3 (84)	23 06·1 (84)	22 56·4 (84)	22 47·1 (84)	22 38·3 (84)
99	23 46·8 (83)	23 35·5 (84)	23 24·7 (83)	23 14·5 (84)	23 04·8 (84)	22 55·5 (84)	22 46·7 (84)
100	23 55·1 (83)	23 43·9 (83)	23 33·0 (84)	23 22·9 (83)	23 13·2 (83)	23 03·9 (83)	22 55·1 (83)
101	24 03·4 (84)	23 52·2 (83)	23 41·4 (83)	23 31·2 (83)	23 21·5 (83)	23 12·2 (83)	23 03·4 (83)
102	24 11·8 (83)	24 00·5 (83)	23 49·7 (83)	23 39·5 (83)	23 29·8 (83)	23 20·5 (83)	23 11·7 (83)
103	24 20·1 (83)	24 08·8 (83)	23 58·0 (83)	23 47·8 (83)	23 38·1	23 28·8	23 20·0
104	24 28·4 (83)	24 17·1 (83)	24 06·3 (83)	23 56·1			
105	24 36·7	24 25·4	24 14·6				
106							
107							

P Upper Branches

Heading of each column group: o, m, s (degrees, minutes, seconds). Sub-branch labels Pd, Pu, Pr appear within the columns as printed.

Surface	0·00	0·01	0·02	0·03	0·04	0·05
Pd	Pd	Pd	Pd	Pd	Pd	Pd
20 4 37·6	4 33·5	19 4 16·9	18 4 1·6	16 3 34·4	14 3 8·1	14 3 6·5
21 49·6	45·4	20 29·2	19 13·6	17 46·5	15 20·1	15 18·2
21·5 55·4	51·2	20·5 35·1	Pu	18 58·3	16 32·1	Pu
Pu	Pu	Pu	19 4 13·9	Pu	17 43·7	15 3 18·5
21·5 4 55·7	4 51·5	20 4 29·5	18 2·4	18 3 58·6	Pu	14 6·7
21 49·9	45·7	19 18·0	17 3 50·8	17 47·1	17 3 43·9	
20 38·4	34·2	18 6·4	Pr	16 35·5	16 32·4	
19 26·8	22·6	17·5 0·6	17 3 50·3	Pr	15 20·7	
18·5 21·0	16·7	Pr	18 4 1·1	17 3 46·0		
Pr	Pr	18 4 5·9				
19 4 26·3	19 4 21·9					

S Upper Branches

Heading of each column group: o, m, s (degrees, minutes, seconds). Sub-branch labels Sd, Su, Sr appear within the columns as printed.

Surface	0·00	0·01	0·02	0·03	0·04
Sd	Sd	Sd	Sd	Sd	Sd
21 8 38·8	8 32·2	20 8 2·8	19 7 35·1	18 7 7·9	17 6 41·8
22 9 0·0	53·2		20 56·3		Su
22·4 8·4	9 1·8	Su	Su	Sr	16 6 21·1
Su	Su	20 8 3·0	20 7 56·4	17 6 47·3	Sr
22 9 0·0	8 53·0	19 7 42·3	19 35·5		16 6 20·8
21 8 39·2	32·1	18 21·8	18 14·9		17 6 40·2
20 18·6	11·4		Sr		
19·6 10·4	3·2	Sr	18 7 14·6		
Sr	Sr	19 7 42·4			
20 8 18·2	10·8				
21 37·4	30·0				

Times of Transmission at Short Epicentral Distances for Different Focal Depths

P

Δ\h/R	0·00	0·002	·004	·006	·008	·010	·012	·014	·016	·018	·020
0·0	5·4	7·0	8·6	10·3	11·9	13·5	15·1	16·7	18·2	19·8	21·
0·1	5·7	7·2	8·8	10·4	12·0	13·6	15·2	16·8	18·3	19·9	21·
0·2	6·6	7·8	9·2	10·7	12·2	13·8	15·4	16·9	18·5	20·1	21·
0·3	7·7	8·8	10·0	11·4	12·8	14·2	15·8	17·2	18·7	20·3	21·8
0·4	9·1	9·8	10·9	12·1	13·4	14·8	16·3	17·7	19·2	20·6	22·
0·5	10·5	11·0	11·9	13·0	14·2	15·6	16·9	18·3	19·7	21·1	22·
0·6	11·9	12·3	13·0	14·1	15·2	16·4	17·6	18·9	20·3	21·7	23·
0·7	13·4	13·6	14·2	15·1	16·2	17·3	18·4	19·6	21·0	22·4	23·
0·8	14·8	15·0	15·5	16·3	17·3	18·3	19·4	20·5	21·7	23·1	24·
0·9	16·2	16·4	16·8	17·5	18·3	19·3	20·4	21·5	22·5	23·8	25·0
1·0	17·7	17·8	18·2	18·8	19·5	20·4	21·4	22·5	23·5	24·7	25·
1·1	19·1	19·2	19·6	20·1	20·7	21·6	22·5	23·5	24·5	25·6	26·
1·2	20·5	20·6	20·9	21·4	22·0	22·9	23·7	24·6	25·5	26·6	27·
1·3	21·9	22·1	22·4	22·8	23·3	24·1	24·8	25·7	26·6	27·6	28·
1·4	23·4	23·5	23·8	24·2	24·7	25·4	26·1	26·8	27·7	28·6	29·
1·5	24·8	25·0	25·3	25·7	26·1	26·7	27·3	28·0	28·8	29·7	30·

S

Δ\h/R	0·00	0·002	·004	·006	·008	·010	·012	·014	·016	·018	·020
0·0	9·2	12·1	15·0	17·8	20·7	23·6	26·4	29·2	32·1	34·9	37·
0·1	9·8	12·4	15·3	18·0	20·9	23·8	26·5	29·3	32·2	35·0	37·
0·2	11·2	13·3	16·0	18·7	21·4	24·2	27·0	29·7	32·6	35·3	38·
0·3	13·2	14·9	17·3	19·8	22·4	25·0	27·6	30·3	33·1	35·8	38·
0·4	15·5	16·7	18·9	21·1	23·5	26·0	28·5	31·1	33·8	36·5	39·
0·5	18·0	18·8	20·6	22·6	24·9	27·2	29·6	32·1	34·7	37·3	40·0
0·6	20·6	21·2	22·6	24·4	26·4	28·6	30·9	33·3	35·8	38·3	40·
0·7	23·1	23·6	24·8	26·4	28·2	30·2	32·3	34·6	37·0	39·4	41·
0·8	25·7	26·0	27·0	28·4	30·1	31·8	33·9	36·0	38·3	40·6	43·
0·9	28·2	28·5	29·4	30·5	32·1	33·7	35·7	37·6	39·8	42·0	44·
1·0	30·8	31·0	31·7	32·8	34·2	35·7	37·5	39·3	41·4	43·5	45·
1·1	33·3	33·5	34·1	35·0	36·3	37·8	39·4	41·1	43·1	45·1	47·
1·2	35·9	36·0	36·5	37·4	38·5	39·9	41·5	43·0	44·9	46·8	48·
1·3	38·4	38·6	39·0	39·7	40·8	42·1	43·5	45·0	46·8	48·6	50·
1·4	41·0	41·1	41·5	42·2	43·2	44·4	45·7	47·1	48·7	50·5	52·
1·5	43·5	43·7	44·1	44·7	45·6	46·6	47·8	49·2	50·7	52·4	54·

$pP - P$

Δ° \ h/R	·00	·01	·02	·03	·04	·05	·06	·07	·08	·09	·10	·11	·12
19	8	17	25										
20	9	19	27										
21	9	20	30	39									
22	9	20	31	41	50								
23	9	21	32	42	51								
24	9	21	32	42	52	61							
25	9	21	32	43	53	62	70						
28	9	22	33	45	55	65	74	81					
29	10	22	33	45	56	65	75	82					
30	10	22	34	45	56	66	75	83	90				
34	10	22	34	46	57	67	77	86	93	100			
35	10	22	34	46	58	68	78	86	94	100			
37	10	22	35	47	58	69	78	87	95	102	107		
40	10	23	35	47	58	69	80	89	97	104	110		
45	10	23	36	48	60	71	81	91	99	107	114	119	124
50	10	23	36	49	61	72	83	93	102	110	118	124	130
55	10	24	37	50	63	74	85	96	105	114	122	129	136
60	10	24	38	51	64	76	88	98	108	117	126	134	141
65	10	24	38	52	65	78	89	100	111	120	129	138	146
70	10	25	39	53	66	79	91	102	113	123	132	141	150
75	10	25	39	53	66	80	92	104	115	125	135	144	153
80	10	25	40	54	67	81	94	105	117	127	137	147	156
85	10	26	40	54	68	82	95	107	118	129	139	149	159
90	10	26	41	55	69	82	96	108	120	131	141	152	162
95	11	26	41	55	69	83	96	108	120	131	142	152	162
100	11	26	41	55	69	83	96	108	140	131	142	153	163

$sS - S$

Δ° \ h/R	·00	·01	·02	·03	·04	·05	·06	·07	·08	·09	·10	·11	·12
19	13d												
20	13d												
21	13d	32d											
22	14d	36d	51d										
23	15r	36r	55r	70d									
24	15	37	57	74r	91								
25	15	37	57	76	94	109							
26	16	37	58	78	96	112	126						
27	16	38	59	78	97	114	130						
28	16	38	59	79	98	116	132						
29	16	38	59	80	99	117	134	148					
30	16	38	60	80	100	118	135	150	162				
35	16	38	60	81	101	120	138	154	168	180	190		
40	16	39	61	83	103	122	141	158	172	185	196	206	
45	16	40	63	84	105	125	145	162	178	191	203	213	222
50	17	41	64	86	107	128	148	166	182	196	210	221	231
55	17	41	65	87	109	130	150	169	186	201	215	228	239
60	17	42	66	89	111	132	153	172	190	206	220	234	246
65	17	42	66	90	113	134	156	176	194	211	226	240	254
70	17	43	67	91	114	137	158	179	198	215	231	246	261
75	17	43	68	93	116	139	161	182	202	220	236	252	268
80	18	44	69	94	118	141	164	185	205	223	241	258	274
85	18	44	70	95	120	143	166	187	208	227	246	263	280
90	18	45	71	96	121	145	169	191	212	231	250	268	286
95	18	45	72	97	122	146	170	193	214	234	253	271	289
100	18	45	72	98	123	147	171	194	215	235	254	272	290

NEAR EARTHQUAKE PHASES

Times of Transmission for a Surface Focus

Δ	Pg	P*	Pn	Sg	S*	Sn
0·0	0 0·0	0 (2·8)	0 (6·8)	0 0·0	0 (3·9)	0 (10·7)
0·2	4·0			6·6		
0·4	8·0	(9·6)		13·2		
0·6	12·0	13·1		19·8	21·7	
0·8	16·0	16·5	18·3	26·4	27·7	31·0
1·0	20·0	19·9	21·1	33·0	33·6	36·1
1·2	23·9	23·3	23·9	39·6	39·5	41·2
1·4	27·9	26·7	26·8	46·3	45·5	46·3
1·6	31·9	30·2	29·6	52·9	51·4	51·3
1·8	35·9	33·6	32·5	59·5	57·4	56·4
2·0	39·9	37·0	35·4	1 6·1	1 3·3	1 1·5
2·2	43·9	40·4	38·3	12·7	9·2	6·6
2·4	47·9	43·8	41·2	19·3	15·2	11·7
2·6	51·9	47·3	44·0	25·9	21·1	16·7
2·8	55·9	50·7	46·9	32·5	27·1	21·8
3·0	59·8	54·1	49·7	39·1	33·0	26·9
3·2	1 3·8	57·5	52·5	45·7	38·9	32·0
3·4	7·8	1 0·9	55·4	52·3	44·9	37·1
3·6	11·8	4·4	58·2	58·9	50·8	42·1
3·8	15·8	7·8	1 1·1	2 5·6	56·8	47·2
4·0	19·8	11·2	3·9	12·2	2 2·7	52·2
4·2	23·8	14·6	6·7	18·8	8·6	57·3
4·4	27·8	18·0	9·6	25·4	14·6	2 2·4
4·6	31·8	21·5	12·4	32·0	20·5	7·4
4·8	35·8	24·9	15·3	38·6	26·5	12·5
5·0	39·8	28·3	18·1	45·2	32·4	17·5
5·2	43·7	31·7	20·9	51·8	38·3	22·5
5·4	47·7	35·1	23·8	58·4	44·3	27·6
5·6	51·7	38·6	26·6	3 5·0	50·2	32·6
5·8	55·7	42·0	29·4	11·6	56·2	37·6
6·0	59·7	45·4	32·2	18·2	3 2·1	42·6
6·2	2 3·7	48·8	35·0	24·8	8·0	47·6
6·4	7·7	52·2	37·9	31·5	14·0	52·6
6·6	11·7	55·7	40·7	38·1	19·9	57·6
6·8	15·7	59·1	43·5	44·7	25·9	3 2·6
7·0	19·6	2 2·5	46·3	51·3	31·8	7·6
7·2	23·6	5·9	49·1	57·9	37·7	12·6
7·4	27·6	9·3	51·9	4 4·5	43·7	17·6
7·6	31·6	12·8	54·7	11·1	49·6	22·6
7·8	35·6	16·2	57·5	17·7	55·6	27·6
8·0	39·6	19·6	2 0·3	24·3	4 1·5	32·6
8·2				30·9	7·4	37·6
8·4				37·5	13·4	42·6
8·6				44·1	19·3	47·6
8·8				50·8	25·3	52·5
9·0				57·4	31·2	57·5
9·2				5 4·0	37·1	4 2·5
9·4				10·6	43·1	7·4
9·6				17·2	49·0	12·4
9·8				23·8	55·0	17·3
10·0				30·4	5 0·9	22·2

Corrections for Focus at the Base of the Upper Layer

Δ	Pg	Sg
0	+2·7	+4·5
0·1	+1·4	+2·3
0·2	+0·8	+1·4
0·3	+0·6	+1·0
0·4	+0·4	+0·7
0·5	+0·4	+0·6
0·6	+0·3	+0·5
0·8	+0·2	+0·4
1·0	+0·2	+0·3
1·2	+0·1	+0·2
1·4	+0·1	+0·2

Δ	P	S
1	−1·9	−2·8
10	−2·0	−3·0
20	−2·2	−3·4
30	−2·4	−3·9
40	−2·5	−4·0
50	−2·5	−4·0
60	−2·5	−4·1
70	−2·5	−4·1
80	−2·6	−4·2
90	−2·6	−4·2
100	−2·6	−4·3
100 (SKS)		−4·4
180 (PKP)	−2·7	
180 (SKS)		4·5

Corrections for Focus at the Base of the Intermediate Layer

Δ	P*	S*
0·0	5·5	9·2
0·1	5·8	9·7
0·2	6·6	11·1
0·3	7·6	13·0
0·4	9·0	15·3
0·5	10·5	18·0
0·6	12·1	20·8
0·7	13·7	23·6
0·8	15·4	26·4

SURFACE REFLEXIONS

Travel-times of the Phases PP, PPP, PS (or SP), PPS (or SPP or PSP), SS, SSP (or PSS or SPS) and SSS. Surface Focus

Δ	PP	PPP	PS	PPS	SS	SSP	SSS
°	m s	m s			m s		m s
0	14$_{143}$	20$_{143}$			21$_{254}$		32$_{254}$
1	28$_{143}$	35$_{143}$			47$_{254}$		58$_{254}$
2	42$_{143}$	49$_{143}$			1 12$_{254}$		1 23$_{254}$
3	57$_{143}$	1 3$_{143}$			38$_{254}$		48$_{254}$
4	1 11$_{143}$	18$_{143}$			2 3$_{254}$		2 14$_{254}$
5	25$_{143}$	32$_{143}$			28$_{254}$		39$_{254}$
6	39$_{143}$	46$_{143}$			54$_{254}$		3 5$_{254}$
7	54$_{142}$	2 1$_{143}$			3 19$_{254}$		30$_{254}$
8	2 8$_{142}$	15$_{143}$			45$_{253}$		55$_{254}$
9	22$_{142}$	29$_{143}$			4 10$_{253}$		4 21$_{254}$
10	36$_{142}$	43$_{142}$			35$_{251}$		46$_{253}$
11	50$_{141}$	58$_{142}$			5 0$_{251}$		5 11$_{253}$
12	3 5$_{141}$	3 12$_{142}$			25$_{251}$		37$_{253}$
13	19$_{140}$	26$_{142}$			50$_{250}$		6 2$_{253}$
14	33$_{140}$	40$_{142}$			6 15$_{250}$		27$_{252}$
15	47$_{140}$	54$_{142}$			40$_{250}$		53$_{251}$
16	4 1$_{139}$	4 9$_{141}$			7 5$_{249}$		7 18$_{251}$
17	15$_{139}$	23$_{141}$			30$_{249}$		43$_{251}$
18	28$_{138}$	37$_{141}$			55$_{248}$		8 8$_{251}$
19	42$_{138}$	51$_{141}$			8 20$_{247}$		33$_{250}$
20	56$_{137}$	5 5$_{140}$			45$_{246}$		58$_{250}$
21	5 10$_{137}$	19$_{140}$			9 9$_{245}$		9 23$_{250}$
22	23$_{136}$	33$_{140}$			34$_{244}$		48$_{250}$
23	37$_{135}$	47$_{140}$			58$_{243}$		10 13$_{250}$
24	51$_{134}$	6 1$_{140}$			10 22$_{242}$		38$_{250}$
25	6 4$_{134}$	15$_{139}$			47$_{241}$		11 3$_{249}$
26	17$_{133}$	29$_{139}$			11 11$_{240}$		28$_{249}$
27	31$_{132}$	43$_{138}$			35$_{239}$		53$_{248}$
28	44$_{131}$	56$_{138}$			59$_{238}$		12 17$_{247}$
29	57$_{131}$	7 10$_{138}$			12 22$_{236}$		42$_{247}$
30	7 10$_{130}$	24$_{137}$			46$_{235}$		13 7$_{246}$
31	23$_{129}$	38$_{137}$			13 9$_{234}$		31$_{245}$
32	36$_{128}$	51$_{137}$			33$_{232}$		56$_{245}$
33	49$_{127}$	8 5$_{137}$			56$_{230}$		14 20$_{244}$
34	8 1$_{125}$	19$_{136}$			14 19$_{229}$		45$_{244}$
35	14$_{125}$	32$_{135}$			42$_{227}$		15 9$_{243}$
36	26$_{124}$	46$_{134}$			15 5$_{227}$		33$_{242}$
37	39$_{122}$	59$_{134}$			27$_{225}$		58$_{242}$
38	51$_{121}$	9 13$_{134}$			50$_{223}$		16 22$_{241}$
39	9 3$_{109}$	26$_{133}$			16 12$_{221}$		46$_{240}$

SURFACE REFLEXIONS—*continued*

Each cell gives the time as m s (minutes shown only when they change), with the interpolation difference printed between rows shown here in parentheses.

Δ	PP	PPP	PS	PPS	SS	SSP	SSS
40	9 14 (104)	9 39 (132)	(13 53)	(14 0)	16 34 (213)		17 10 (239)
41	24 (104)	53 (132)			56 (193)		34 (239)
42	35 (102)	10 6 (131)			17 15 (190)		58 (238)
43	45 (100)	19 (131)			34 (186)		18 22 (237)
44	55 (99)	32 (131)	14 50 (143)	14 57 (143)	52 (183)		45 (236)
45	10 5 (99)	45 (130)	15 5 (143)	15 12 (143)	18 11 (180)		19 9 (235)
46	15 (98)	58 (130)	19 (143)	26 (143)	29 (177)		32 (234)
47	25 (97)	11 11 (129)	33 (143)	40 (143)	46 (174)		56 (234)
48	34 (96)	24 (128)	48 (143)	54 (143)	19 4 (171)		20 19 (233)
49	44 (96)	37 (127)	16 2 (142)	16 9 (143)	21 (169)		42 (231)
50	54 (95)	49 (127)	16 (142)	23 (143)	38 (168)		21 6 (230)
51	11 3 (93)	12 2 (126)	30 (142)	37 (143)	55 (166)		29 (229)
52	12 (93)	15 (125)	45 (142)	52 (142)	20 11 (165)		51 (228)
53	22 (92)	27 (124)	59 (141)	17 6 (142)	28 (164)		22 14 (227)
54	31 (91)	40 (124)	17 13 (141)	20 (142)	44 (162)		37 (227)
55	40 (91)	52 (123)	27 (141)	34 (142)	21 0 (161)		23 0 (226)
56	49 (90)	13 4 (122)	41 (140)	48 (142)	16 (161)		22 (225)
57	58 (90)	17 (122)	55 (139)	18 3 (142)	33 (160)		45 (223)
58	12 7 (90)	29 (117)	18 9 (139)	17 (141)	49 (160)		24 7 (222)
59	16 (89)	40 (106)	23 (138)	31 (141)	22 5 (159)		29 (221)
60	25 (89)	51 (105)	37 (138)	45 (141)	20 (159)		51 (217)
61	34 (88)	14 2 (104)	50 (138)	59 (141)	36 (159)		25 13 (200)
62	43 (88)	12 (103)	19 4 (137)	19 13 (141)	52 (157)		33 (192)
63	51 (88)	22 (102)	18 (136)	27 (140)	23 8 (156)		52 (190)
64	13 0 (88)	32 (101)	32 (136)	41 (140)	23 (156)		26 11 (188)
65	9 (87)	43 (100)	45 (135)	55 (139)	39 (156)		30 (186)
66	18 (87)	53 (100)	59 (135)	20 9 (139)	54 (155)		49 (183)
67	26 (86)	15 3 (99)	20 12 (134)	23 (139)	24 10 (155)		27 7 (182)
68	35 (86)	12 (98)	26 (133)	37 (138)	25 (155)		25 (180)
69	44 (86)	22 (98)	39 (133)	51 (138)	41 (155)		43 (177)
70	52 (85)	32 (97)	52 (132)	21 5 (138)	56 (155)		28 1 (175)
71	14 1 (85)	42 (97)	21 5 (132)	18 (138)	25 12 (154)		18 (173)
72	9 (85)	51 (97)	19 (131)	32 (138)	27 (154)		36 (172)
73	18 (84)	16 1 (96)	32 (131)	46 (137)	43 (154)		53 (170)
74	26 (84)	11 (96)	45 (130)	22 0 (136)	58 (153)		29 10 (169)
75	35 (84)	20 (95)	58 (129)	13 (136)	26 13 (152)		27 (168)
76	43 (84)	30 (94)	22 11 (128)	27 (136)	29 (152)		44 (167)
77	51 (83)	39 (94)	23 (128)	40 (135)	44 (152)		30 0 (166)
78	15 0 (83)	49 (93)	36 (127)	54 (135)	59 (151)		17 (165)
79	8 (83)	58 (92)	49 (127)	23 7 (134)	27 14 (151)		33 (164)
80	16 (83)	17 7 (91)	23 1 (125)	21 (133)	29 (150)		50 (163)
81	25 (82)	16 (91)	14 (124)	34 (133)	44 (149)		31 6 (163)
82	33 (82)	25 (91)	26 (124)	47 (133)	59 (148)		22 (162)
83	41 (82)	34 (91)	39 (123)	24 1 (133)	28 14 (148)		39 (161)
84	49 (82)	44 (90)	51 (123)	14 (132)	28 (147)		55 (161)

SURFACE REFLEXIONS—*continued*

Δ	PP	PPP	PS	PPS	SS	SSP	SSS
85	15 57 (81)	17 53 (90)	24 3 (122)	24 27 (132)	28 43 (146)		32 11 (161)
86	16 5 (81)	18 2 (90)	16 (122)	40 (131)	58 (146)		27 (160)
87	14 (81)	11 (90)	28 (120)	54 (131)	29 12 (145)		43 (160)
88	22 (81)	20 (90)	40 (119)	25 7 (131)	27 (145)		59 (159)
89	30 (80)	29 (89)	52 (119)	20 (130)	41 (144)		33 15 (159)
90	38 (80)	37 (89)	25 3 (118)	33 (129)	56 (144)	30 3 (143)	31 (159)
91	46 (80)	46 (88)	15 (117)	46 (129)	30 10 (143)	17 (143)	47 (158)
92	54 (79)	55 (88)	27 (113)	59 (128)	24 (143)	31 (143)	34 2 (157)
93	17 2 (79)	19 4 (88)	38 (110)	26 11 (128)	39 (142)	46 (143)	18 (156)
94	10 (78)	13 (88)	49 (108)	24 (127)	53 (142)	31 0 (143)	34 (156)
95	17 (78)	22 (88)	26 0 (108)	37 (127)	31 7 (141)	14 (143)	49 (156)
96	25 (78)	30 (88)	11 (107)	50 (126)	21 (141)	28 (142)	35 5 (156)
97	33 (77)	39 (87)	22 (107)	27 2 (125)	35 (140)	43 (142)	20 (156)
98	41 (77)	48 (87)	32 (106)	15 (125)	49 (140)	57 (142)	36 (156)
99	48 (77)	57 (87)	43 (105)	27 (124)	32 3 (139)	32 11 (142)	52 (156)
100	56 (77)	20 5 (87)	53 (105)	40 (124)	17 (138)	25 (141)	36 7 (155)
101	18 4 (76)	14 (86)	27 4 (104)	52 (123)	31 (138)	39 (141)	23 (155)
102	11 (76)	23 (86)	14 (103)	28 4 (123)	45 (138)	53 (141)	38 (155)
103	19 (75)	31 (86)	25 (102)	17 (123)	59 (137)	33 8 (141)	54 (155)
104	26 (75)	40 (86)	35 (101)	29 (122)	33 12 (137)	22 (140)	37 9 (155)
105	34 (75)	48 (85)	45 (100)	41 (122)	26 (136)	36 (140)	25 (155)
106	41 (74)	57 (85)	55 (99)	53 (121)	40 (136)	50 (140)	40 (155)
107	49 (74)	21 5 (85)	28 5 (98)	29 5 (120)	53 (135)	34 4 (139)	56 (154)
108	56 (73)	14 (84)	15 (98)	17 (120)	34 7 (135)	18 (139)	38 11 (154)
109	19 4 (73)	22 (84)	24 (97)	29 (119)	20 (134)	31 (138)	26 (153)
110	11 (73)	31 (84)	34 (97)	41 (118)	34 (134)	45 (138)	42 (153)
111	18 (72)	39 (84)	44 (96)	53 (117)	47 (133)	59 (138)	57 (153)
112	25 (72)	47 (84)	53 (94)	30 5 (110)	35 0 (133)	35 13 (137)	39 12 (153)
113	33 (71)	56 (84)	29 3 (94)	16 (108)	14 (132)	26 (137)	28 (152)
114	40 (71)	22 4 (84)	12 (93)	27 (107)	27 (132)	40 (136)	43 (152)
115	47 (70)	13 (84)	22 (92)	37 (107)	40 (132)	53 (136)	58 (151)
116	54 (70)	21 (84)	31 (92)	48 (107)	53 (131)	36 6 (135)	40 13 (151)
117	20 1 (69)	30 (83)	40 (91)	59 (107)	36 6 (131)	20 (135)	28 (151)
118	8 (69)	38 (83)	49 (90)	31 9 (106)	19 (130)	33 (134)	43 (151)
119	15 (69)	46 (83)	58 (90)	20 (105)	32 (129)	47 (134)	58 (151)
120	21 (68)	54 (82)	30 7 (89)	30 (105)	45 (128)	37 0 (134)	41 13 (150)
121	28 (68)	23 3 (82)	16 (89)	41 (104)	58 (128)	14 (133)	28 (150)
122	35 (68)	11 (82)	25 (88)	51 (103)	37 11 (127)	27 (133)	43 (149)
123	42 (67)	19 (82)	34 (88)	32 2 (103)	24 (127)	41 (133)	58 (148)
124	49 (67)	27 (82)	42 (87)	12 (102)	36 (126)	54 (132)	42 13 (148)
125	55 (66)	35 (82)	51 (87)	22 (101)	49 (126)	38 8 (132)	28 (148)
126	21 2 (66)	44 (82)	31 0 (87)	32 (101)	38 1 (125)	21 (132)	43 (147)
127	8 (66)	52 (82)	9 (87)	42 (100)	14 (124)	35 (132)	57 (147)
128	15 (65)	24 0 (82)	17 (86)	52 (100)	26 (124)	48 (132)	43 12 (146)
129	22 (65)	8 (81)	26 (85)	33 2 (99)	39 (123)	39 2 (132)	27 (146)

SURFACE REFLEXIONS—continued   SURFACE REFLEXIONS—continued

Δ (°)	PP (m s)	PPP (m s)	PS (m s)	PPS (m s)	SS (m s)	SSP (m s)	SSS (m s)
130	21 28 (65)	24 16 (81)	31 34 (85)	33 12 (98)	38 51 (123)	39 15 (131)	43 41 (145)
131	35 (64)	24 (81)	43 (85)	22 (98)	39 3 (122)	28 (131)	56 (145)
132	41 (64)	32 (81)	51 (85)	32 (97)	16 (122)	41 (130)	44 10 (145)
133	47 (63)	41 (81)	32 0 (84)	42 (97)	28 (121)	54 (130)	25 (145)
134	54 (63)	49 (80)	8 (84)	51 (97)	40 (120)	40 7 (130)	39 (144)
135	22 0 (63)	57 (80)	17 (84)	34 1 (96)	52 (120)	20 (129)	54 (144)
136	6 (62)	25 5 (80)	25 (84)	11 (96)	40 4 (119)	33 (128)	45 8 (144)
137	12 (62)	13 (79)	33 (84)	20 (95)	16 (119)	46 (128)	22 (143)
138	19 (62)	20 (79)	42 (84)	30 (95)	28 (119)	59 (128)	37 (143)
139	25 (62)	28 (79)	50 (84)	39 (94)	39 (118)	41 12 (127)	51 (143)
140	31 (61)	36 (79)	59 (84)	48 (93)	51 (117)	24 (127)	46 5 (142)
141	37 (60)	44 (79)	33 7 (83)	58 (92)	41 3 (116)	37 (126)	19 (142)
142	43 (60)	52 (79)	15 (83)	35 7 (92)	15 (116)	50 (126)	34 (141)
143	49 (60)	26 0 (78)	24 (83)	16 (92)	26 (115)	42 2 (125)	48 (141)
144	55 (59)	8 (78)	32 (83)	25 (91)	38 (114)	15 (125)	47 2 (141)
145	23 1 (59)	16 (77)	40 (83)	34 (90)	49 (113)	27 (125)	16 (141)
146	7 (59)	23 (77)	49 (83)	44 (90)	42 0 (113)	40 (124)	30 (140)
147	13 (58)	31 (77)	57	53 (90)	12 (112)	52 (124)	44 (140)
148	19 (58)	39 (77)		36 2 (90)	23 (112)	43 4 (123)	58 (140)
149	24 (57)	46 (77)		11 (90)	34 (111)	17 (123)	48 12 (139)
150	30 (57)	54 (76)		20 (89)	45 (111)	29 (122)	26 (138)
151	36 (57)	27 2 (76)		28 (89)	56 (111)	41 (122)	40 (138)
152	41 (56)	9 (76)		37 (88)	43 7 (110)	53 (121)	53 (138)
153	47 (56)	17 (76)		46 (88)	18 (109)	44 5 (121)	49 7 (138)
154	53 (56)	24 (76)		55 (88)	29 (109)	17 (120)	21 (138)
155	58 (55)	32 (76)		37 4 (88)	40 (108)	29 (120)	35 (137)
156	24 4 (55)	40 (75)		13 (87)	51 (108)	41 (119)	49 (137)
157	9 (54)	47 (75)		21 (87)	44 1 (107)	53 (119)	50 2 (136)
158	15 (54)	55 (74)		30 (87)	12 (106)	45 5 (119)	16 (136)
159	20 (54)	28 2 (74)		39 (87)	23 (106)	17 (119)	29 (136)
160	25 (53)	9 (74)		47 (87)	33 (105)	29 (118)	43 (136)
161	31 (53)	17 (74)		56 (87)	43 (104)	41 (117)	57 (136)
162	36 (53)	24 (74)		38 5 (86)	54 (104)	53 (116)	51 10 (135)
163	41 (53)	32 (73)		13 (86)	45 4 (103)	46 4 (116)	24 (135)
164	47 (52)	39 (73)		22 (85)	15 (102)	15 (113)	37 (134)
165	52 (51)	46 (73)		30 (85)	25 (101)	27 (112)	50 (134)
166	57 (51)	53 (73)		39 (85)	35 (100)	38 (112)	52 4 (134)
167	25 2 (51)	29 1 (72)		47 (85)	45 (100)	49 (112)	17 (134)
168	7 (51)	8 (72)		56 (85)	55 (99)	47 0 (111)	31 (133)
169	12 (51)	15 (72)		39 4 (84)	46 5 (98)	11 (111)	44 (133)
170	17 (50)	22 (71)		13 (84)	15 (98)	23 (110)	57 (132)
171	22 (49)	29 (71)		21 (84)	24 (97)	34 (109)	53 10 (132)
172	27 (49)	36 (70)		30 (84)	34 (96)	44 (109)	23 (132)
173	32 (49)	43 (70)		38 (84)	44 (95)	55 (108)	37 (132)
174	37 (49)	50 (70)		46 (84)	53 (95)	48 6 (108)	50 (131)
175	25 42 (48)	29 57 (70)		39 55 (84)	47 3 (94)	48 17 (107)	54 3 (131)
176	47 (48)	30 4 (70)		40 3 (84)	12 (93)	28 (107)	16 (131)
177	51 (47)	11 (69)		12 (84)	21 (93)	38 (107)	29 (130)
178	56 (47)	18 (69)		20 (84)	31 (92)	49 (106)	42 (130)
179	26 1 (47)	25 (69)		28 (84)	40 (91)	49 0 (105)	55 (129)
180	5 (46)	32 (68)		37 (83)	49 (90)	10 (104)	55 8 (128)
181	10 (46)	39 (68)		45 (83)	58 (89)	20 (104)	21 (128)
182	15 (46)	46 (68)		53 (83)	48 7 (89)	31 (104)	33 (128)
183	19 (46)	53 (68)		41 2 (83)	16 (88)	41 (104)	46 (128)
184	24 (46)	59 (68)		10 (83)	25 (88)	52 (103)	59 (127)
185	28 (46)	31 6 (67)		18 (83)	33 (87)	50 2 (103)	56 12 (127)
186	33 (46)	13 (67)		27 (83)	42 (87)	12 (102)	24 (127)
187	38 (46)	20 (66)		35 (83)	51 (86)	22 (102)	37 (126)
188	42 (46)	26 (66)		43 (83)	59 (86)	32 (101)	50 (126)
189	47 (46)	33 (66)			49 8 (85)	42 (100)	57 2 (126)
190	51 (46)	39 (66)			16 (85)	52 (100)	15 (125)
191	56 (46)	46 (66)			25 (85)	51 2 (99)	27 (125)
192	27 1 (46)	53 (65)			33 (85)	12 (98)	40 (124)
193	5 (45)	59 (65)			42 (85)	22 (98)	52 (124)
194	10 (45)	32 6 (65)			50 (84)	32 (98)	58 4 (123)
195	14 (45)	12 (64)			59 (84)	41 (98)	17 (123)
196	19 (45)	18 (64)			50 7 (84)	51 (97)	29 (123)
197	23 (45)	25 (64)			16 (84)	52 1 (97)	41 (122)
198	28 (45)	31 (64)			24 (84)	10 (96)	53 (122)
199	32 (45)	38 (64)			32 (84)	20 (95)	59 6 (122)
200	37 (45)	44 (64)			41 (84)	29 (95)	17 (121)
201	41 (45)	50 (63)			49 (84)	39 (94)	29 (121)
202	46 (45)	57 (63)			58 (84)	48 (94)	41 (120)
203	50 (45)	33 3 (63)			51 6 (84)	58 (93)	53 (120)
204	55 (44)	9 (62)			14 (83)	53 7 (93)	60 6 (120)
205	59 (44)	16 (62)			23 (83)	16 (92)	18 (119)
206	28 4 (44)	22 (62)			31 (83)	25 (92)	30 (119)
207	8 (44)	28 (61)			39 (83)	34 (91)	41 (119)
208	12 (44)	34 (61)			48 (83)	43 (91)	53 (118)
209	17 (44)	40 (61)			56 (83)	52 (90)	61 5 (118)
210	21	46 (61)			52 4 (83)	54 1 (90)	17 (118)
211		52 (61)			13 (83)	10 (90)	29 (117)
212		58 (61)			21 (83)	19 (89)	40 (116)
213		34 5 (60)			29 (83)	28 (89)	52 (116)
214		11 (60)			37	37 (88)	62 4 (115)
215		17 (60)				46 (88)	15 (114)
216		23 (59)				55 (88)	26 (114)
217		28 (59)				55 3 (87)	38 (114)
218		34 (59)				12 (87)	49 (113)
219		40 (58)				21 (87)	63 1 (113)
220		46				30	12

Depth Allowances. PP

Δ \ h	·00	·01	·02	·03	·04	·05	·06	·07	·08	·09	·10	·11	·12
10	3	3											
20	4	5	6	7									
30	4	7	10	12	13	14	13	14	14				
32	4	7	10	13	14	16	17	19	20				
34	4	8	11	14	15	18	21	24	25	25			
36	4	8	11	14	18	22	25	28	30	31			
38	4	8	12	17	22	26	29	32	34	36	36		
40	4	10	15	20	24	28	32	35	38	39	40	40	
45	5	10	16	21	26	30	34	38	41	43	45	46	46
50	5	10	16	22	27	32	36	40	43	46	48	50	51
60	5	11	17	23	28	33	38	42	46	49	52	54	55
70	5	11	17	23	29	34	39	44	48	51	54	57	59
80	5	11	18	24	29	35	40	45	49	53	56	59	61
90	5	12	18	24	30	36	41	46	50	54	57	61	64
100	5	12	18	25	31	36	42	47	51	55	59	63	66
110	5	12	19	25	32	37	43	48	53	57	61	65	69
120	5	12	19	26	32	38	44	49	54	59	63	67	71
130	5	12	19	26	33	39	45	50	56	60	65	69	74
140	5	12	19	26	33	40	46	51	57	62	66	71	75
150	5	13	20	27	33	40	46	52	58	63	68	72	77
160	5	13	20	27	34	40	47	53	59	64	69	74	79
170	5	13	20	27	34	41	47	54	59	65	70	75	80
180	5	13	20	28	35	41	48	54	60	65	71	76	81

Depth Allowances. PPP

Δ \ h	·00	·01	·02	·03	·04	·05	·06	·07	·08	·09	·10	·11	·12
10	3												
20	4	4											
30	4	5	6										
40	4	7	9	9	9								
50	4	8	11	13	15	15	15	17					
52	4	8	11	14	16	17	18	21	21				
54	4	8	11	14	16	19	22	25	26				
56	4	8	12	15	19	23	26	29	30	31			
58	4	9	14	18	22	26	30	33	34	35			
60	4	10	15	20	24	28	32	35	37	38	39		
70	5	10	16	21	26	30	35	38	41	43	45	47	
80	5	11	17	22	28	32	37	41	44	47	49	51	52
90	5	11	17	23	28	33	38	42	46	49	51	53	55
100	5	11	17	23	29	34	39	43	47	50	53	55	57
120	5	11	17	24	29	35	40	44	49	52	55	58	61
140	5	11	18	24	30	36	41	46	50	54	58	61	64
160	5	12	18	25	31	37	42	47	52	56	60	64	68
180	5	12	19	26	32	38	44	49	54	59	63	67	71
200	5	12	19	26	33	39	45	51	56	61	65	70	75
220	5	12	19	26	33	40	46	52	57	62	67	72	77

Depth Allowances. PS and SP

(a) PS

Δ \ h	·00	·01	·02	·03	·04	·05	·06	·07	·08	·09	·10	·11	·12
50	3	3											
60	3	5	6										
70	4	7	9	10	10								
80	4	8	11	14	15	16	17						
85	4	8	11	15	17	19	21	22					
90	4	9	12	16	21	23	26	28	30				
95	4	9	14	19	23	26	30	32	34	34			
100	5	10	15	20	24	28	32	35	37	38	39		
105	5	10	16	21	25	30	34	37	40	42	43	44	44
110	5	11	16	21	26	31	36	39	42	45	47	48	49
120	5	11	17	22	28	33	38	42	46	49	51	54	55
130	5	11	17	23	29	34	39	44	47	51	54	57	58
140	5	11	17	23	29	35	40	44	48	52	55	58	59

(b) SP

Δ \ h	·00	·01	·02	·03	·04	·05	·06	·07	·08	·09	·10	·11	·12
40	8	20	31	42	53	63	73	81	89	96	102	108	113
50	8	20	31	42	53	63	73	82	90	97	103	109	114
60	8	20	32	43	54	64	74	83	92	99	106	112	118
70	8	21	32	44	55	66	76	85	94	102	109	116	122
80	8	21	33	45	56	67	77	87	96	105	112	120	127
85	9	21	33	45	57	68	78	88	98	106	114	122	129
90	9	21	34	46	57	68	79	89	99	108	116	124	133
95	9	22	34	47	58	70	81	92	102	111	119	128	136
100	9	22	34	47	59	70	82	93	103	112	121	129	139
105	9	22	35	47	60	71	83	94	104	113	122	131	140
110	9	22	35	48	60	72	83	94	105	114	124	133	141
120	9	22	35	48	61	73	84	95	106	116	126	135	143
130	9	22	36	49	61	73	85	96	107	117	127	136	145
140	9	23	36	49	61	73	85	97	107	118	127	137	145

Depth Allowances. PPS, PSP and SPP

(a) PPS or PSP

$\Delta \backslash h$	·00	·01	·02	·03	·04	·05	·06	·07	·08	·09	·10	·11	·12
60	3	3											
70	4	5											
80	4	6	8	8									
90	4	7	10	11	12								
100	4	8	11	14	16	17	18	(17)					
110	4	9	12	18	21	24	27	29	28				
120	4	10	15	20	24	28	31	34	36	37			
130	5	10	16	21	26	30	34	38	40	42	44	44	
140	5	11	16	22	27	32	36	40	43	46	49	50	51
160	5	11	17	23	28	34	38	43	47	50	53	55	57
180	5	11	17	23	29	35	40	44	48	52	55	(58)	(60

(b) SPP

$\Delta \backslash h$	·00	·01	·02	·03	·04	·05	·06	·07	·08	·09	·10	·11	·12
40	8	20	31	42	53	63	73	81	89	96	102	108	113
50	8	20	31	42	53	63	73	81	89	97	103	109	113
60	8	20	32	43	53	63	73	82	90	98	104	110	115
70	8	20	32	43	54	64	74	83	92	99	106	112	117
80	8	21	32	43	54	65	75	84	93	101	108	115	120
90	8	21	32	44	55	66	77	86	95	103	111	118	124
100	8	21	33	45	56	67	78	88	97	105	113	121	127
110	9	22	34	46	58	69	80	91	101	110	118	126	134
120	9	22	34	47	59	69	82	92	103	112	121	129	137
140	9	22	35	48	60	72	84	95	105	115	124	133	142
160	9	22	36	49	61	73	85	96	107	117	126	135	144
180	9	22	36	49	61	73	85	97	107	118	127	137	145

Depth Allowances. SS

$\Delta \backslash h$	·00	·01	·02	·03	·04	·05	·06	·07	·08	·09	·10	·11	·12
10	5	6											
20	6	9	10	(10)									
30	6	11	15	18	20	20	18	19	18				
32	6	12	16	20	22	24	24	26	27	25			
34	6	12	17	21	25	27	30	34	36	36	35		
36	6	13	18	23	27	32	37	42	45	46	47		
38	6	13	19	25	32	39	45	50	53	56	58	58	
40	6	14	23	31	39	46	52	58	62	66	68	69	
42·5	7	17	26	34	43	50	57	64	69	73	76	78	78
45	8	18	26	35	45	53	61	68	74	78	83	84	85
50	8	19	29	39	48	56	65	72	78	84	88	91	93
60	8	19	30	40	50	59	68	76	83	89	94	97	100
70	8	19	30	40	51	60	69	77	84	91	96	100	104
80	8	20	31	41	52	61	71	79	87	93	99	105	109
90	8	20	31	42	53	63	72	81	89	96	103	108	113
100	8	20	32	43	54	64	74	83	91	99	106	112	117
110	8	20	32	44	55	65	75	85	93	101	108	115	121
120	8	21	33	44	55	66	77	86	95	103	111	118	124
130	8	21	33	45	56	67	78	88	97	106	114	121	128
140	9	21	34	46	57	68	79	90	99	108	116	124	131
150	9	21	34	47	58	69	81	91	101	110	119	127	135
160	9	22	35	47	59	71	82	93	103	112	121	129	138
170	9	22	35	48	60	72	83	94	104	114	123	132	140
180	9	22	35	48	61	73	84	96	106	116	126	135	143

Depth Allowances. PSS, SSP and SPS

(a) PSS

$\Delta \backslash h$	·00	·01	·02	·03	·04	·05	·06	·07	·08	·09	·10	·11	·12
100	3	3											
110	4	5											
120	4	6	7										
130	4	7	9	10									
140	4	7	10	13	14								
150	4	8	11	15	17	18	19						
160	4	8	12	16	19	22	24	26					
170	4	9	14	18	22	26	28	31	31				
180	4	10	15	20	24	28	31	34	36	37			
200	5	10	16	22	27	31	36	40	43	46	48	49	
220	5	11	17	23	28	34	39	43	47	50	53	55	57

(b) *SSP or SPS*

Δ \ h	·00	·01	·02	·03	·04	·05	·06	·07	·08	·09	·10	·11	·12
90	8	20	31	42	53	63	72	81	88	96	103	108	113
100	8	20	32	43	53	63	73	82	90	97	104	110	115
110	8	20	32	43	54	64	74	83	91	99	106	112	117
120	8	20	32	43	54	65	75	84	93	101	108	114	120
130	8	21	33	44	55	66	76	85	94	102	110	116	122
140	8	21	33	44	56	67	77	87	96	104	112	119	125
150	8	21	33	45	56	68	78	88	97	106	114	121	128
160	9	21	33	45	57	68	79	89	99	108	116	124	131
170	9	21	34	46	58	69	81	91	101	110	119	127	135
180	9	22	35	47	59	71	82	93	103	112	121	129	138
200	9	22	35	48	60	72	84	95	105	115	124	133	142
220	9	22	36	48	61	73	85	96	107	117	126	136	144

Depth Allowances. SSS

Δ \ h	·00	·01	·02	·03	·04	·05	·06	·07	·08	·09	·10	·11	·12
10	5												
20	5	6											
30	6	8	9										
40	6	10	13	14									
50	6	12	17	20	22	23		20					
52	6	12	17	21	24	25		27	27				
54	6	12	18	22	25	28	30	34	35				
56	6	13	18	23	27	31	37	42	43	43			
58	6	13	19	25	31	38	43	48	51	52	52		
60	7	15	22	30	37	44	50	55	59	61	62		
62	7	17	25	33	41	48	55	60	64	67	69		
64	7	17	26	34	42	50	57	63	67	71	73	73	
66	8	17	27	36	44	52	59	65	70	74	77	78	
68	8	18	27	36	45	53	60	67	73	78	80	82	
70	8	18	28	37	46	54	62	69	75	81	83	85	86
80	8	19	29	39	49	58	66	74	80	85	90	93	95
90	8	19	30	40	50	59	68	76	83	88	93	97	100
100	8	19	30	40	50	60	69	77	84	90	95	99	102
120	8	20	31	41	52	61	71	79	86	93	99	104	108
140	8	20	32	43	53	63	73	82	90	97	104	109	114
160	8	20	32	43	54	65	75	84	92	100	107	113	119
180	8	21	33	44	55	66	77	86	95	103	111	118	124
200	9	21	33	45	57	68	78	88	98	106	114	122	129
220	9	21	34	46	58	69	80	91	100	109	118	120	133

CORE REFLEXIONS

PcP. Surface Focus

Δ	m	s	Δ	m	s	Δ	m	s	Δ	m	s
0	8	34·3	26	9	5·2	51	10	22·0	76	12	3·4
1		34·3	27		7·5	52		25·7	77		7·7
2		34·5	28		9·9	53		29·4	78		12·0
3		34·7	29		12·4	54		33·2	79		16·3
4		35·0	30		14·9	55		37·0	80		20·6
5		35·4	31		17·5	56		40·9	81		24·9
6		35·9	32		20·2	57		44·8	82		29·3
7		36·6	33		22·9	58		48·7	83		33·6
8		37·3	34		25·7	59		52·6	84		37·9
9		38·1	35		28·6	60		56·6	85		42·3
10		39·0	36		31·5	61	11	0·6	86		46·7
11		40·0	37		34·5	62		4·6	87		51·0
12		41·1	38		37·6	63		8·7	88		55·4
13		42·3	39		40·7	64		12·8	89		59·8
14		43·5	40		43·9	65		16·9	90	13	4·2
15		44·9	41		47·1	66		21·0	91		8·6
16		46·4	42		50·4	67		25·2	92		13·0
17		47·9	43		53·7	68		29·4	93		17·4
18		49·5	44		57·1	69		33·6	94		21·9
19		51·2	45	10	0·5	70		37·8	95		26·3
20		53·0	46		4·0	71		42·0	96		30·7
21		54·9	47		7·5	72		46·3	97		35·2
22		56·8	48		11·1	73		50·5	98		39·6
23		58·8	49		14·7	74		54·8	99		44·1
24	9	0·9	50		18·3	75		59·1	100		48·5
25		3·0									

ScS. Surface Focus

Δ	m	s	Δ	m	s	Δ	m	s	Δ	m	s
0	15	35·7	26	16	33·0	51	18	54·6	76	22	3·1
1		35·8	27		37·3	52	19	1·5	77		11·1
2		36·1	28		41·7	53		8·4	78		19·2
3		36·5	29		46·3	54		15·4	79		27·4
4		37·1	30		51·0	55		22·5	80		35·5
5		37·9	31		55·8	56		29·6	81		43·7
6		38·9	32	17	0·8	57		36·8	82		51·9
7		40·1	33		5·9	58		44·1	83	23	0·1
8		41·4	34		11·1	59		51·4	84		8·3
9		42·9	35		16·4	60		58·8	85		16·5
10		44·6	36		21·8	61	20	6·2	86		24·7
11		46·5	37		27·4	62		13·7	87		33·0
12		48·5	38		33·0	63		21·2	88		41·3
13		50·7	39		38·8	64		28·8	89		49·5
14		53·0	40		44·6	65		36·4	90		57·8
15		55·5	41		50·5	66		44·1	91	24	6·1
16		58·2	42		56·5	67		51·8	92		14·3
17	16	1·0	43	18	2·6	68		59·6	93		22·6
18		3·9	44		8·8	69	21	7·4	94		30·9
19		7·0	45		15·1	70		15·2	95		39·2
20		10·3	46		21·5	71		23·1	96		47·5
21		13·7	47		28·0	72		31·0	97		55·8
22		17·3	48		34·5	73		39·0	98	25	4·1
23		21·0	49		41·1	74		47·0	99		12·4
24		24·9	50		47·8	75		55·0	100		20·7
25		28·9									

Depth Allowances. PcP

h / Δ	·00	·01	·02	·03	·04	·05	·06	·07	·08	·09	·10	·11	·12
0	−5·4	−13·5	−21·4	−29·1	−36·6	−43·9	−51·1	−58·0	−64·5	−70·8	−76·8	−82·8	−88·8
10	−5·4	−13·4	−21·3	−29·0	−36·5	−43·8	−50·9	−57·8	−64·3	−70·5	−76·5	−82·5	−88·5
20	−5·4	−13·4	−21·2	−28·8	−36·2	−43·5	−50·6	−57·4	−63·8	−69·9	−75·8	−81·6	−87·4
30	−5·3	−13·3	−21·0	−28·6	−35·9	−43·0	−50·0	−56·7	−63·1	−69·1	−74·8	−80·5	−86·2
40	−5·3	−13·2	−20·9	−28·4	−35·6	−42·6	−49·5	−56·1	−62·4	−68·3	−73·9	−79·5	−84·9
50	−5·3	−13·1	−20·7	−28·1	−35·2	−42·2	−49·0	−55·5	−61·7	−67·4	−72·9	−78·2	−83·3
60	−5·3	−13·1	−20·6	−27·9	−35·0	−41·9	−48·6	−55·0	−61·1	−66·8	−72·3	−77·6	−82·8
70	−5·3	−13·0	−20·5	−27·7	−34·8	−41·6	−48·3	−54·7	−60·7	−66·3	−71·7	−77·0	−82·2
80	−5·3	−13·0	−20·5	−27·7	−34·8	−41·6	−48·3	−54·6	−60·5	−66·1	−71·5	−76·8	−81·9

Depth Allowances. ScS

h / Δ	·00	·01	·02	·03	·04	·05	·06	·07	·08	·09	·10	·11	·12
0	−9·2	−23·6	−37·7	−51·5	−65·1	−78·3	−91·2	−103·6	−115·5	−126·9	−137·9	−148·6	−159·0
10	−9·2	−23·5	−37·6	−51·4	−64·9	−78·1	−90·9	−103·2	−115·1	−126·4	−137·4	−148·1	−158·4
20	−9·2	−23·4	−37·4	−51·1	−64·5	−77·6	−90·3	−102·6	−114·3	−125·4	−136·2	−146·7	−156·9
30	−9·1	−23·3	−37·1	−50·7	−63·9	−76·8	−89·4	−101·2	−112·7	−123·7	−134·4	−144·7	−154·6
40	−9·1	−23·1	−36·8	−50·2	−63·3	−76·0	−88·3	−100·1	−111·4	−122·3	−132·7	−142·8	−152·4
50	−9·1	−23·0	−36·5	−49·7	−62·6	−75·1	−87·3	−99·0	−110·1	−120·7	−130·9	−140·7	−150·1
60	−9·0	−22·8	−36·3	−49·4	−62·2	−74·6	−86·7	−98·2	−109·1	−119·5	−129·5	−139·1	−148·3
70	−9·0	−22·6	−36·0	−49·1	−61·8	−74·1	−85·9	−97·4	−108·3	−118·6	−128·4	−137·8	−146·8
80	−9·0	−22·6	−35·9	−48·9	−61·5	−73·7	−85·5	−96·9	−107·7	−117·8	−127·5	−136·8	−145·7

PcS = ScP. Surface Focus

Δ	m	s		Δ	m	s
0	12	5·0		36	13	17·8
1		5·1		37		21·5
2		5·2		38		25·2
3		5·5		39		29·0
4		6·0		40		32·9
5		6·5		41		36·8
6		7·2		42		40·8
7		8·0		43		44·8
8		9·0		44		48·8
9		10·0		45		52·9
10		11·2		46		57·0
11		12·5		47	14	1·2
12		13·9		48		5·4
13		15·4		49		9·6
14		17·1		50		13·8
15		18·8		51		18·1
16		20·6		52		22·4
17		22·6		53		26·7
18		24·7		54		31·0
19		26·8		55		35·4
20		29·1		56		39·8
21		31·5		57		44·2
22		33·9		58		48·6
23		36·5		59		53·0
24		39·1		60		57·4
25		41·9		61	15	1·9
26		44·8		62		6·3
27		47·7		63		10·7
28		50·8		64		15·2
29		53·9		65		19·6
30		57·1		66		24·1
31	13	0·4		67		28·5
32		3·7		68		32·9
33		7·2		69		37·3
34		10·7		70		41·7
35		14·2				

Partition of Distance

Δ	Δ (as P)	Δ (as S)
0	0	0
5	3	2
10	6	4
15	10	5
20	14	6
25	17	8
30	20	10
35	24	11
40	28	12
45	32	13
50	37	13
55	42	13
60	46	14
65	51	14
70	56	14

ScSP

Δ	m	s		Δ	m	s
140	32	59		113	29	18
130	31	36		114		24
120	30	14		115		29
115	29	34		120		54
112		12		125	30	17
112		12		130		39

Surface Focus

PKP

Δ	m	s
180A	22	10·6
179		6·2
178		1·8
177	21	57·4
176		52·9
175		48·5
174		44·1
173		39·6
172		35·2
171		30·8
170		26·3
169		21·9
168		17·5
167		13·1
166		8·8
165		4·4
164		0·1
163	20	55·7
162		51·4
161		47·0
160		42·7
159		38·4
158		34·1
157		29·8
156		25·5
155		21·2
154		17·0
153		12·8
152		8·6

Δ	m	s
151	20	4·4
150		0·2
149	19	56·1
148		52·0
147		48·1
146		44·2
145		40·4
144		36·8
143B		33·5
143B		33·5
144		36·5
145		39·3
146		41·8
147C		44·1
110D	18	33·2
111		35·2
112		37·1
113		39·1
114		41·1
115		43·0
116		45·0
117		46·9
118		48·8
119		50·8
120		52·7
121		54·7
122		56·6

Δ	m	s
123	18	58·5
124	19	0·5
125		2·4
126		4·3
127		6·2
128		8·2
129		10·1
130		12·0
131		13·9
132		15·8
133		17·7
134		19·5
135		21·4
136		23·2
137		25·0
138		26·9
139		28·7
140		30·5
141		32·3
142		34·0
143		35·7
144		37·4
145E		39·2
146		40·9
147		42·6
148		44·2
149		45·8
150		47·4
151		48·9

Δ	m	s
152	19	50·4
153		51·8
154		53·2
155		54·5
156		55·8
157		57·2
158		58·5
159		59·7
160	20	0·8
161		1·8
162		2·8
163		3·8
164		4·8
165		5·8
166		6·7
167		7·4
168		8·0
169		8·6
170		9·2
171		9·8
172		10·4
173		10·9
174		11·2
175		11·5
176		11·8
177		12·0
178		12·1
179		12·1
180F		12·2

PKS

Δ	m	s
148A	23	55·0
145		41·8
142		28·4
141		23·9
140		19·5
139		15·0
138		10·6
137		6·2
136		1·9
135	22	57·6
134		53·2
133		48·9
132		44·6
131		40·4
130B		36·2
130B		36·2
131		39·9
132		43·4
133		46·6
134		49·7
135		52·7
136		55·6
137		58·3
138	23	0·9
139		3·4
140C		5·7

Δ	m	s
104D	21	57·0
105		59·0
106	22	0·9
107		2·9
108		4·8
109		6·8
110		8·7
111		10·7
112		12·6
113		14·6
114		16·5
115		18·4
116		20·4
117		22·3
118		24·3
119		26·2
120		28·1
121		30·1
122		32·0
123		34·0
124		35·9
125		37·7
126		39·6
127		41·5
128		43·4
129		45·2

Δ	m	s
130	22	47·0
131		48·9
132		50·7
133		52·6
134		54·4
135		56·3
136		58·1
137		59·8
138	23	1·6
139E		3·3
140		5·0
141		6·7
142		8·3
143		10·0
144		11·7
145		13·3
146		14·9
147		16·4
148		17·9
149		19·3
150		20·8
151		22·2
152		23·5
153		24·8
154		26·1
155		27·4

Δ	m	s
156	23	28·6
157		29·7
158		30·8
159		31·8
160		32·8
161		33·8
162		34·8
163		35·8
164		36·6
165		37·3
166		37·9
167		38·6
168		39·2
169		39·8
170		40·4
171		41·0
172		41·4
173		41·7
174		42·0
175		42·3
176		42·5
177		42·7
178		42·8
179		42·9
180F		42·9

SKS

Δ	m	s
62A	20	13·7
63		21·2
64		28·7
65		36·2
66		43·7
67		51·1
68		58·6
69	21	6·0
70		13·4
71		20·8
72		28·2
73		35·5
74		42·9
75		50·2
76		57·5
77	22	4·8
78		12·0
79		19·1
80		26·3
81		33·4
82		40·4
83		47·3
84		54·1
85	23	0·8
86		7·5
87		14·1
88		20·5
89		26·7
90		32·8
91		38·7
92		44·4
93		50·1
94		55·6
95	24	1·1
96		6·5
97		11·8
98		17·0
99		22·1
100		27·0

Δ	m	s
101	24	31·9
102		36·7
103		41·4
104		46·0
105		50·5
106		55·0
107		59·4
108	25	3·7
109		8·0
110		12·2
111		16·3
112		20·3
113		24·3
114		28·2
115		32·0
116		35·7
117		39·4
118		43·0
119		46·5
120		50·0
121		53·3
122		56·6
123		59·8
124	26	2·9
125		5·9
126		8·9
127		11·8
128		14·6
129		17·3
130		19·9
131		22·4
132		24·9
133C		27·3
99D	25	22·9
100		24·9
101		26·8
102		28·8
103		30·7

Δ	m	s
104	25	32·7
105		34·6
106		36·6
107		38·5
108		40·5
109		42·4
110		44·3
111		46·3
112		48·2
113		50·2
114		52·1
115		54·0
116		56·0
117		57·9
118		59·9
119	26	1·8
120		3·6
121		5·5
122		7·3
123		9·2
124		11·1
125		13·0
126		14·8
127		16·7
128		18·5
129		20·3
130		22·1
131		23·9
132		25·7
133E		27·4
134		29·1
135		30·8
136		32·5
137		34·2
138		35·9
139		37·5
140		39·1
141		40·8
142		42·4

Δ	m	s
143	26	43·9
144		45·4
145		46·9
146		48·3
147		49·7
148		51·1
149		52·5
150		53·8
151		55·1
152		56·3
153		57·5
154		58·6
155		59·7
156	27	0·7
157		1·7
158		2·7
159		3·7
160		4·7
161		5·6
162		6·4
163		7·1
164		7·8
165		8·5
166		9·2
167		9·8
168		10·4
169		11·0
170		11·5
171		11·9
172		12·3
173		12·6
174		12·9
175		13·1
176		13·3
177		13·4
178		13·4
179		13·5
180F		13·5

PKKP

Δ	m	s	Δ	m	s	Δ	m	s	Δ	m	s
256A	30	15·1	234B	28	39·4	257	29	55·7	250	29	50
255		10·7	235		43·4	258		58·4	255		59
254		6·2	236		47·4	259	30	1·1	260	30	8
253		1·8	237		51·4	260		3·7	265		17
252	29	57·4	238		55·1	261		6·2	270		25
251		52·9	239		58·8	262		8·7	275		34
250		48·5	240	29	2·5	263		11·2	280		42
249		44·0	241		6·1	264		13·7	285		50
248		39·6	242		9·6	265		16·1	290		57
247		35·2	243		13·1	266		18·4	295	31	4
246		30·9	244		16·4	267C		20·7	300		11
245		26·5	245		19·7				305		17
244		22·2	246		23·0	198D	28	10	310		23
243		17·8	247		26·3	200		14	315		28
242		13·5	248		29·4	205		23	320		32
241		9·2	249		32·5	210		33	325		37
240		4·8	250		35·5	215		43	330		40
239		0·5	251		38·5	220		52	335		43
238	28	56·2	252		41·5	225	29	2	340		46
237		52·0	253		44·4	230		12	345		48
236		47·8	254		47·3	235		22	350		49
235		43·6	255		50·1	240		31	355		50
234B		39·4	256		52·9	245		40	360F		50

PKKS

Δ	m	s	Δ	m	s	Δ	m	s	Δ	m	s
226A	32	8·3	221	31	43·4	241	32	51·7	192D	31	33
225		3·9	222		47·2	242		54·7	200		49
224	31	59·5	223		50·9	243		57·6	210	32	8
223		55·1	224		54·6	244	33	0·6	220		28
222		50·7	225		58·3	245		3·5	230		47
221		46·2	226	32	2·0	246		6·3	240	33	6
220		41·8	227		5·6	247		9·1	250		25
219		37·3	228		9·2	248		11·9	260		43
218		32·9	229		12·8	249		14·6	270	34	0
217		28·5	230		16·3	250		17·3	280		16
216		24·0	231		19·7	251		19·9	290		30
215		19·6	232		23·1	252		22·5	300		44
214B		15·2	233		26·4	253		25·1	310		55
214B		15·2	234		29·7	254		27·7	320	35	4
215		19·5	235		33·0	255		30·3	330		11
216		23·6	236		36·3	256		32·8	340		17
217		27·6	237		39·5	257		35·3	350		20
218		31·6	238		42·6	258		37·7	360F		21
219		35·6	239		45·7	259		40·0			
220		39·5	240		48·7	260C		42·3			

SKKS

Δ	m	s	Δ	m	s	Δ	m	s	Δ	m	s
85A	23	5	125	27	47	165	31	38	205	34	35
86		12	126		54	166		43	206		39
87		20	127	28	0	167		48	207		43
88		27	128		7	168		53	208		47
89		34	129		13	169		58	209		50
90		42	130		20	170	32	3	210		54
91		49	131		26	171		8	211		58
92		56	132		32	172		13	212	35	2
93	24	4	133		38	173		18	213		5
94		11	134		45	174		22	214		9
95		18	135		51	175		27	215		12
96		25	136		57	176		32	216		16
97		33	137	29	3	177		36	217		19
98		40	138		9	178		41	218		23
99		47	139		15	179		46	219		26
100		54	140		21	180		50	220		30
101	25	1	141		27	181		55	221		33
102		9	142		33	182		59	222		36
103		16	143		39	183	33	4	223		40
104		23	144		44	184		8	224		43
105		30	145		50	185		13	225		46
106		37	146		56	186		17	226		49
107		44	147	30	2	187		22	227		52
108		51	148		7	188		26	228		56
109		58	149		13	189		30	229		59
110	26	5	150		18	190		34	230	36	2
111		12	151		24	191		39	231		5
112		19	152		29	192		43	232		8
113		26	153		35	193		47	233		11
114		33	154		40	194		51	234		14
115		40	155		46	195		56	235		17
116		47	156		51	196	34	0	236		20
117		54	157		57	197		4	237		23
118	27	0	158	31	2	198		8	238		26
119		7	159		7	199		12	239		28
120		14	160		12	200		16	240		31
121		20	161		18	201		20	241		34
122		27	162		23	202		24	242		36
123		34	163		28	203		28	243C		39
124		41	164		33	204		32			

SKKS (DF)

Δ	m	s	Δ	m	s	Δ	m	s	Δ	m	s
187	34	59	230	36	23	275	37	42	320	38	36
190	35	5	235		32	280		49	325		39
195		15	240		41	285		57	330		43
200		25	245		50	290	38	4	335		45
205		34	250	37	0	295		10	340		48
210		44	255		9	300		16	345		50
215		54	260		17	305		21	350		51
220	36	4	265		26	310		27	355		51
225		13	270		34	315		32	360		52

SKKKS (ABC)

Δ	m	s	Δ	m	s	Δ	m	s	Δ	m	s
90	23	43	165	32	33	235	39	15	305	44	16
95	24	20	170	33	6	240		39	310		34
100		57	175		38	245	40	3	315		52
105	25	33	180	34	9	250		27	320	45	9
110	26	10	185		39	255		50	325		26
115		46	190	35	9	260	41	13	330		42
120	27	22	195		38	265		35	335		58
125		58	200	36	7	270		57	340	46	13
130	28	34	205		36	275	42	18	345		28
135	29	9	210	37	3	280		39	350		42
140		44	215		30	285		59	355		56
145	30	19	220		57	290	43	19	360	47	9
150		53	225	38	24	295		39	365		21
155	31	27	230		50	300		58	370		34
160	32	0									

PcPPKP

Δ	m	s	Δ	m	s	Δ	m	s	Δ	m	s
278A	35	50·3	180	29	8·0	142	27	49·5	162	28	26·9
275		37·0	179		5·1	143		51·4	163		28·6
270		14·8	178		2·2	144		53·4	164		30·3
265	34	52·6	177	28	59·5	145		55·3	165		32·0
260		30·5	176		56·8	146		57·3	166		33·7
255		8·5	175		54·2	147		59·2	167		35·3
250	33	46·7	174		51·7	148	28	1·1	168		36·8
245		25·0	173B		49·3	149		3·1	169		38·1
240		3·3	174C		51·6	150		5·0	170		39·4
235	32	41·8				151		7·0	171		40·6
230		20·3	132D	27	30·0	152		8·9	172		41·7
225	31	59·1	133		32·0	153		10·8	173		42·8
220		38·0	134		33·9	154		12·6	174		43·8
215		17·1	135		35·9	155		14·5	175		44·6
210	30	56·6	136		37·8	156		16·3	176		45·3
205		36·8	137		39·8	157		18·1	177		45·8
200		17·4	138		41·7	158		19·9	178		46·3
195	29	58·7	139		43·7	159		21·7	179		46·5
190		40·8	140		45·6	160		23·5	180F		46·5
185		23·8	141		47·6	161		25·2			

PcSPKP

Δ	m	s	Δ	m	s	Δ	m	s	Δ	m	s
250A	37	52	210	34	57	170	32	21	150	31	40
245		30	205		35	166C		11	155		49
240		8	200		14				160		58
235	36	46	195	33	53	127D	30	56	165	32	6
230		24	190		33	130	31	2	170		12
225		2	185		13	135		11	175		16
220	35	40	180	32	55	140		21	180F		17
215		18	175		37	145		31			

ScSPKP

Δ	m	s	Δ	m	s	Δ	m	s	Δ	m	s
206A	38	43·5	179	36	45·8	127	34	31·4	154	35	22·0
205		39·1	178		41·6	128		33·4	155		23·7
204		34·6	177		37·4	129		35·3	156		25·4
203		30·2	176		33·2	130		37·3	157		27·1
202		25·8	175		29·1	131		39·2	158		28·7
201		21·3	174		25·0	132		41·1	159		30·3
200		16·8	173		20·9	133		43·1	160		31·8
199		12·4	172		16·9	134		45·0	161		33·3
198		8·0	171		13·0	135		47·0	162		34·7
197		3·6	170		9·1	136		48·9	163		36·0
196	37	59·2	169		5·2	137		50·8	164		37·3
195		54·8	168		1·4	138		52·8	165		38·6
194		50·4	167	35	57·7	139		54·7	166		39·7
193		46·1	166		54·1	140		56·7	167		40·7
192		41·7	165		50·5	141		58·6	168		41·7
191		37·4	164		47·1	142	35	0·5	169		42·7
190		33·0	163		43·7	143		2·4	170		43·7
189		28·7	162		40·5	144		4·2	171		44·5
188		24·3	161		37·5	145		6·1	172		45·2
187		20·0	160BC		34·8	146		7·9	173		45·8
186		15·7				147		9·7	174		46·4
185		11·4	121D	34	19·7	148		11·5	175		46·9
184		7·1	122		21·7	149		13·3	176		47·3
183		2·8	123		23·6	150		15·1	177		47·6
182	36	58·5	124		25·6	151		16·9	178		47·8
181		54·3	125		27·5	152		18·6	179		47·9
180		50·0	126		29·5	153		20·3	180		47·9

SKSP

Δ	m	s	Δ	m	s	Δ	m	s	Δ	m	s
115	29	34	140	32	37	165	35	18	190	37	30
120	30	12	145	33	12	170		47	195		53
125		49	150		45	175	36	14	200	38	16
130	31	26	155	34	17	180		40	205		39
135	32	2	160		48	185	37	6			

PKPPKP

Δ	m	s	Δ	m	s	Δ	m	s
286B	39	7	255	38	14	310	39	49
288		13	260		24	315		56
290		19	265		34	320	40	1
			270		43	325		7
220D	37	6	275		52	330		12
225		16	280	39	1	335		15
230		26	285		10	340		18
235		36	290		18	345		21
240		45	295		27	350		23
245		55	300		35	355		24
250	38	5	305		42	360F		24

PKPPKS

Δ	m	s	Δ	m	s	Δ	m	s	Δ	m	s
276B	42	20	215D	40	32	265	42	8	315	43	28
277		23	220		42	270		18	320		34
278		26	225		52	275		27	325		39
279		29	230	41	2	280		36	330		43
280		32	235		11	285		44	335		47
281		35	240		21	290		52	340		50
282		38	245		31	295	43	1	345		52
283		40	250		40	300		8	350		54
284		43	255		50	305		15	355		55
285		45	260		59	310		22	360F		55

PKP. Depth Allowances

(To be subtracted from times for surface focus)

Δ	·00	·01	·02	·03	·04	·05	·06	·07	·08	·09	·10	·11	·12
142	5·3	13·2	20·9	28·4	35·6	42·6	49·5	56·1	62·4	68·4	74·1	79·8	85·3
145	5·3	13·2	21·0	28·5	35·9	43·1	50·1	56·8	63·2	69·3	75·2	81·0	86·7
150	5·4	13·4	21·3	28·9	36·4	43·7	50·8	57·6	64·1	70·2	76·2	82·2	88·1
160	5·4	13·4	21·3	29·0	36·5	43·8	50·9	57·8	64·3	70·4	76·4	82·4	88·4
170	5·4	13·5	21·4	29·1	36·6	43·9	51·1	58·0	64·5	70·7	76·7	82·7	88·7
180	5·4	13·5	21·4	29·1	36·6	43·9	51·1	58·0	64·5	70·8	76·8	82·8	88·8

$PKP_2 = PKP$ (AB)

Δ													
160	5·3	12·9	20·4	27·7	34·8	41·7	48·3	54·8	60·8	66·3	71·5	76·6	81·7

SKS. Depth Allowances

(To be subtracted from times for surface focus)

Δ	·00	·01	·02	·03	·04	·05	·06	·07	·08	·09	·10	·11	·12
70	9·0	22·7	36·1	49·3	62·2	74·7	86·7	98·1	109·1	119·6	129·7	139·4	148·6
80	9·0	22·8	36·3	49·6	62·5	75·0	87·1	98·7	109·9	120·5	130·7	140·5	150·0
85	9·0	22·9	36·5	49·8	62·7	75·3	87·5	99·1	110·3	121·0	131·3	141·2	150·7
90	9·1	23·0	36·7	50·1	63·2	75·9	88·1	99·9	111·3	122·0	132·6	142·6	152·2
95	9·1	23·1	36·8	50·4	63·6	76·4	88·8	100·7	112·1	123·0	133·7	143·9	153·7
100	9·1	23·1	37·0	50·5	63·8	76·7	89·3	101·2	112·6	123·7	134·5	144·9	154·8
105	9·1	23·2	37·1	50·7	64·0	77·0	89·6	101·6	113·1	124·2	135·0	145·4	155·4
110	9·1	23·3	37·2	50·9	64·2	77·2	89·8	101·9	113·5	124·7	135·5	145·9	156·0
115	9·2	23·4	37·4	51·1	64·5	77·5	90·1	102·2	113·9	125·2	136·0	146·5	156·6
120	9·2	23·5	37·5	51·2	64·6	77·6	90·3	102·5	114·2	125·5	136·4	146·9	157·1
125	9·2	23·5	37·5	51·2	64·6	77·7	90·5	102·7	114·4	125·7	136·6	147·2	157·5
130	9·2	23·5	37·6	51·3	64·7	77·9	90·7	102·9	114·7	126·0	136·9	147·5	157·8
135	9·2	23·5	37·6	51·4	64·9	78·1	90·9	103·1	115·0	126·3	137·2	147·9	158·3
140	9·2	23·5	37·6	51·4	65·0	78·2	91·0	103·3	115·2	126·5	137·5	148·2	158·6
145	9·2	23·5	37·6	51·4	65·0	78·2	91·0	103·3	115·3	126·6	137·6	148·3	158·7
150	9·2	23·5	37·6	51·5	65·1	78·2	91·1	103·4	115·3	126·6	137·6	148·3	158·8
180	9·2	23·6	37·7	51·6	65·2	78·3	91·2	103·6	115·5	126·8	137·8	148·5	159·0

Times within the Core

K (ABE)

Δ	m	s	Δ	m	s	Δ	m	s	Δ	m	s
0	0	0·0	31	3	45·8	61	6	50·1	91	9	7·2
1		7·5	32		52·7	62		55·4	92		11·0
2		15·0	33		59·5	63	7	0·7	93		14·7
3		22·5	34	4	6·3	64		5·9	94		18·4
4		30·0	35		13·0	65		11·1	95		22·1
5		37·4	36		19·7	66		16·2	96		25·7
6		44·9	37		26·4	67		21·3	97		29·3
7		52·3	38		33·0	68		26·3	98		32·8
8		59·8	39		39·6	69		31·2	99		36·3
9	1	7·2	40		46·1	70		36·1	100		39·7
10		14·6	41		52·6	71		40·9	101		43·0
11		22·0	42		59·0	72		45·7	102		46·3
12		29·4	43	5	5·4	73		50·4	103		49·6
13		36·8	44		11·7	74		55·1	104		52·8
14		44·1	45		17·9	75		59·7	105		55·9
15		51·5	46		24·1	76	8	4·3	106		59·0
16		58·8	47		30·2	77		8·8	107	10	2·0
17	2	6·1	48		36·3	78		13·3	108		5·0
18		13·4	49		42·3	79		17·7	109		7·9
19		20·7	50		48·2	80		22·1	110		10·8
20		27·9	51		54·1	81		26·4	111		13·6
21		35·1	52		59·9	82		30·7	112		16·4
22		42·3	53	6	5·7	83		35·0	113		19·1
23		49·5	54		11·4	84		39·2	114		21·8
24		56·7	55		17·1	85		43·3	115		24·4
25	3	3·8	56		22·7	86		47·4	116		26·9
26		10·9	57		28·3	87		51·4	117		29·4
27		18·0	58		33·8	88		55·4	118		31·9
28		25·0	59		39·3	89		59·4	119		34·3
29		32·0	60		44·7	90	9	3·3	120		36·6
30		38·9									

$KIK = K$ (DF)

Δ	m	s	Δ	m	s	Δ	m	s	Δ	m	s
88	9	36·4	112	10	22·8	135	11	2·5	158	11	29·4
89		38·4	113		24·6	136		4·0	159		30·1
90		40·3	114		26·5	137		5·4	160		30·8
91		42·3	115		28·4	138		6·9	161		31·5
92		44·2	116		30·2	139		8·3	162		32·1
93		46·2	117		32·1	140		9·7	163		32·7
94		48·1	118		33·9	141		11·0	164		33·3
95		50·1	119		35·7	142		12·3	165		33·9
96		52·0	120		37·5	143		13·6	166		34·5
97		54·0	121		39·3	144		14·9	167		35·0
98		55·9	122		41·1	145		16·2	168		35·5
99		57·8	123		42·8	146		17·4	169		35·9
100		59·8	124		44·5	147		18·6	170		36·2
101	10	1·7	125		46·2	148		19·7	171		36·5
102		3·7	126		47·9	149		20·8	172		36·8
103		5·6	127		49·6	150		21·8	173		37·1
104		7·6	128		51·3	151		22·8	174		37·3
105		9·5	129		53·0	152		23·8	175		37·5
106		11·5	130		54·6	153		24·8	176		37·6
107		13·4	131		56·2	154		25·8	177		37·7
108		15·3	132		57·9	155		26·7	178		37·8
109		17·2	133		59·5	156		27·7	179		37·9
110		19·0	134	11	1·0	157		28·6	180		37·9
111		20·9									

I

Δ	m	s	Δ	m	s	Δ	m	s	Δ	m	s
0	0	0·0	50	1	34·6	95	2	44·4	140	3	29·0
5		9·8	55		43·3	100		50·8	145		32·0
10		19·5	60		51·8	105		56·8	150		34·7
15		29·2	65	2	0·1	110	3	2·5	155		37·0
20		38·9	70		8·2	115		7·8	160		38·8
25		48·5	75		16·0	120		12·8	165		40·3
30		58·0	80		23·5	125		17·4	170		41·3
35	1	7·3	85		30·8	130		21·6	175		41·9
40		16·6	90		37·8	135		25·5	180		42·1
45		25·7									

sP−P

h/R \ Δ	15 km.	·00	·01	·02	·03	·04	·05	·06	·07	·08	·09	·10	·11	·12
1	6	12	21											
5	6	12	24	35	45	54	61	68	73	78				
10	6	12	26	39	52	63	73	83	92	100	107	112	117	120
15	7	12	28	43	57	70	84	97	109	119	128	135	142	147
20	7	13	32	49	66	82	97	112	125	137	148	157	166	172
25	7	13	33	51	70	87	104	120	135	149	161	173	184	193
30	7	14	33	52	71	89	106	123	138	152	166	178	189	200
35	7	14	34	53	72	90	108	124	140	155	168	181	193	204
40	7	14	34	54	73	91	108	125	142	156	170	183	196	207
50	7	14	34	54	74	93	111	128	145	160	175	189	202	214
60	7	14	35	55	75	95	113	131	148	164	180	194	208	222
70	7	14	35	56	76	96	115	134	152	168	184	199	213	228
80	7	14	36	57	77	97	117	136	154	171	187	203	218	232
90	7	14	36	58	78	98	118	137	156	173	190	205	221	236
100	7	14	36	58	78	98	118	138	156	173	190	206	222	237

sP Depth Allowances

h/R \ Δ	15 km.	·00	·01	·02	·03	·04	·05	·06	·07	·08	·09	·10	·11	·12
1	4	8	20											
5	4	8	20	32	43	53	63	73	82	90				
10	4	8	20	32	43	54	64	74	83	91	98	104	110	114
15	4	8	21	33	44	55	66	76	85	93	101	107	113	118
20	4	9r	22r	34r	46d	57	68	79	88	98	106	113	120	125
25	4	9	22	35	48	60	72	83	94	104	114	123	132	140
30	4	9	22	36	48	61	73	84	96	106	116	125	134	143
35	4	9	22	36	49	61	73	85	96	107	117	126	136	144
40	4	9	22	36	49	61	74	85	97	107	118	127	136	145
50	4	9	23	36	49	62	74	86	98	109	119	129	138	147
60	4	9	23	36	50	63	75	87	99	110	120	131	140	150
70	4	9	23	37	50	63	76	88	100	111	122	132	142	152
80	4	9	23	37	50	64	76	89	101	112	123	134	144	154
90	4	9	23	37	51	64	77	90	102	113	124	135	145	155
100	4	9	23	37	51	64	77	90	102	113	124	135	145	155
sP begins at	1°	1°	2°	2°	3°	3°	4°	4°	5°	6°	7°	8°	9°	
Overtakes S at	1°	1°·5	2°·3	3°·2	3°·9	4°·6	5°·5	6°·4	7°·4	8°·6	10°·1	12°·0	13°·8	

sPKP Depth Allowances

h/R \ Δ	15 km.	·00	·01	·02	·03	·04	·05	·06	·07	·08	·09	·10	·11	·12
180 A*	4	9	24	38	51	64	77	90	102	113	124	135	146	156
150	4	9	24	38	51	64	77	90	102	114	125	136	146	156
143–4 B	4	9	24	38	52	65	78	91	103	114	126	137	147	157
110 D	4	9	24	38	52	65	78	91	103	115	126	137	·148	158
140	4	9	24	38	52	65	78	91	103	115	126	137	148	159
180 F	4	9	24	38	52	65	78	91	104	116	127	138	149	159

pPKP−PKP, sSKS−SKS

These are twice the depth allowances for PKP and SKS respectively, within 1ˢ.

sPcP, sPKS Depth Allowances

These may be taken to be the same as for sPKP at a distance where $dt/d\Delta$ is the same.

Dr. Harold Jeffreys and Dr. K. E. Bullen,

PKP. Depth Allowances between 110° *and* 140°

h/R	0·00	0·01	0·02	0·03	0·04	0·05	0·06	0·07	0·08	0·09	0·10	0·11	0·12
	5·4	13·4	21·2	28·8	36·2	43·5	50·5	57·2	63·7	69·8	75·7	81·5	87·3

SKS − P

(For rapid determination of focal depth when pP, sS are not available)

h/R Δ	Surface	0·00	0·01	0·02	0·03	0·04	0·05	0·06	0·07	0·08	0·09	0·10	0·11	0·12
	m s	s	s	s	s	m s	m s	s	s	s	s	s	s	s
85°	10 22	18	12	6	0	9 54	9 48	42	37	31	26	21	16	12
90°	30	26	20	14	8	10 1	9 56	50	44	39	34	28	24	19
95°	35	32	25	19	12	6	10 0	54	49	43	38	32	28	23
100°	39	35	28	22	16	9	3	57	52	46	41	35	30	25

HEIGHT ABOVE MEAN SPHERE

φ	h
° ′	
0 00	
	+ 7
10 11	
	+ 6
16 12	
	+ 5
20 42	
	+ 4
24 29	
	+ 3
27 53	
	+ 2
31 00	
	+ 1
33 58	
	0
36 48	
	− 1
39 33	
	− 2
42 15	
	− 3
44 55	
	− 4
47 37	
	− 5
50 17	
	− 6
53 01	
	− 7
55 50	
	− 8
58 46	
	− 9
61 51	
	− 10
65 11	
	− 11
68 52	
	− 12
73 11	
	− 13
78 50	
	− 14
90 00	

In critical cases ascend.

TABLES OF $f(\Delta)$

Correction $= f(\Delta)\,(h + H)$

P		S		SKS	
Δ	$f(\Delta)$	Δ	$f(\Delta)$	Δ	$f(\Delta)$
°	s	°	s	°	s
9		9		79	
	0·01		0·01		0·11
13		10		83	
	0·02		0·02		0·12
22		13		106	
	0·03		0·03		0·13
41		16		116	
	0·04		0·04		0·14
63		22		124	
	0·05		0·05		0·15
72		32		140	
	0·06		0·06		
83		44			
	0·07		0·07		
124		54			
	0·08		0·08		
140		66			
			0·09		
		71			
			0·10		
		76			
			0·11		
		82			
			0·12		
		103			
			0·13		
In critical		123		For *PKP*	
cases ascend.			0·14	$f(\Delta) = 0^{s}·09$	
		140			